Lecture Note
in Control an

Editors: M. Thoma · M. Morari

Springer

*Berlin
Heidelberg
New York
Hong Kong
London
Milan
Paris
Tokyo*

M. Nakamura · S. Goto · N. Kyura

Mechatronic Servo System Control

Problems in Industries and their Theoretical Solutions

Translated by Tao Zhang

With 79 Figures and 21 Tables

Springer

Series Advisory Board

A. Bensoussan · P. Fleming · M.J. Grimble · P. Kokotovic ·
A.B. Kurzhanski · H. Kwakernaak · J.N. Tsitsiklis

Authors

Masatoshi Nakamura
Department of Advanced Systems
Control Engineering
Saga University
Japan

Satoru Goto
Department of Advanced Systems
Control Engineering
Saga University
Japan

Nobuhiro Kyura
Department of Electrical and
Communication Engineering
Kyushu School of Engineering
Kinki University
Japan

Translator:
Tao Zhang
Intelligent Systems Research Division
National Institute of Informatics
Japan

Translation from the Japanese edition
© Morikita Shuppan Co., Ltd. 1998
All Rights Reserved.

ISSN 0170-8643

ISBN 3-540-21096-2   Springer-Verlag Berlin Heidelberg New York

Library of Congress Control Number: 2004103117

This work is subject to copyright. All rights are reserved, whether the whole or part of the material is concerned, specifically the rights of translation, reprinting, reuse of illustrations, recitation, broadcasting, reproduction on microfilm or in other ways, and storage in data banks. Duplication of this publication or parts thereof is permitted only under the provisions of the German Copyright Law of September 9, 1965, in its current version, and permission for use must always be obtained from Springer-Verlag. Violations are liable for prosecution under German Copyright Law.

Springer-Verlag is a part of Springer Science+Business Media

springeronline.com

© Springer-Verlag Berlin Heidelberg 2004
Printed in Germany

The use of general descriptive names, registered names, trademarks, etc. in this publication does not imply, even in the absence of a specific statement, that such names are exempt from the relevant protective laws and regulations and therefore free for general use.

Typesetting: Data conversion by the authors.
Final processing by PTP-Berlin Protago-TeX-Production GmbH, Berlin
Cover-Design: design & production GmbH, Heidelberg
Printed on acid-free paper   62/3020Yu - 5 4 3 2 1 0

From the Main Author

As a main author of Mechatronic Servo System Control (in Japanese), I would like to express my thanks to Dr. Zhang Tao who translated our book into English. The authors, myself, Dr. Goto and Prof. Kyura published the original book which mainly consisted of the authors' original research achievement of mechatronic servo systems control during last over ten years. The original book was fortunately awarded as the best book from the Society of Instrument and Control Engineering (SICE in Japan) in 2001. Moreover, the book was already translated into Korea language by a Korean publisher. As the authors believe that our book is effective for students and engineers who are involved in the field of Mechatronic Control and Robotics, we have been intended the translation of it in English. The authors themselves made the Japanese-English dictionary for the terminologies in the book, and ask to Dr. Zhang Tao for the translation of the book by use of the dictionary. Dr. Zhang Tao has completed the translation by use of his every night times during last several months. I would like to show my great gratitude for his effort for the translation. I also express my great thanks to Prof. Jeffrey Johnson and Dr. Mike Richards (Open University in UK) who helped the final check of the translation.

November 2002   Main author Masatoshi NAKAMURA

From the Translator

Since the term "Mechatronics" was first introduced by a Yaskawa Electric engineer in 1969, and its rigorous definition was given by a technical committee, i.e., The International Federation for the Theory of Machines and Mechanism (IFToMM), as "*Mechatronics is the synergistic combination of precision mechanical engineering, electronic control, and systems thinking in design of products and manufacturing processes*", the development of mechatronic techniques has led to widespread adoption of electronics in machinery. At the same time, as one of the key techniques of mechatronics, servo control system has been well defined for various kinds of mechanical system. At present, mechatronic techniques are essential for advanced mechanical engineering. Furthermore, the introduction of servo control system design to engineers engaged in mechanical engineering is thought to be indispensable.

As a researcher on mechatronic technique, when I firstly read the Japanese version of this book "Mechatronic Servo System Control", written by Prof. Nakamura, Dr. Goto and Prof. Kyura, I was attracted by its meticulous study on the issues of mechatronic servo control system arising from mechanical engineering as well as the significance of application. Additionally, I aroused a strong desire to transfer its valuable achievements to whole researchers and engineers who are engaging in the mechatronic techniques or willing to obtain knowledge related with mechatronic techniques. After I heard that this book was awarded the 2001 Works Reward of The Society of Instrument and Control Engineers (SICE), and Prof. Nakamura also had the same desire to translate it into other languages for readers, I expressed my strong wish to be responsible for translating this book into English. With deep trust and great encouragement from Prof. Nakamura, I started this challenging project from one year ago.

Through the great efforts, the English version of "Mechatronic Servo Control System" was finished recently. As I read the English version of this book once again, I have also obtained great enlightenment from it, particularly for my further research on mechatronic techniques. From the contents of this book, I believe all readers will share the same feeling. The profit of this book will be reflected not only in the research or teaching on mechatronic techniques, but also for engineers working on mechanical engineering.

Finally, I also want to express my great gratitude to Prof. Nakamura, Dr. Goto and Prof. Kyura to distribute such a great valuable book on their achievements within several decades of years to whole readers. For the kindly help from Dr. D. Kushida during the period of my translation, especially the valuable review of this book from Prof. Jeffrey Johnson and Dr. Mike Richards, Open University, UK, I transfer my deep appreciate to them.

Because of my insufficiency of knowledge on translation between Japanese and English, there might have some mistakes in this book. It will be very kind if you can indicate them to me and I will make my best efforts continuously to improve them.

November 2002   Tao ZHANG

Preface

The editor and composer is engaging in the study on systems control and their applications in university. As one of his research fields, with a plenty of opportunities of discussion with Kyura, who is long-term working on the servo controller design and its application in mechatronic industry, on the control of mechatronic machine during the past ten years, the cooperative research has been made greatly progress. These discussion meetings were held several times once a year. Achievements on the items of these discussion meetings were compiled into reports, each of which has between 50~100 pages. Then, many valuable commends were obtained from Kyura in terms of these achievements. Moreover, new research directions were found. The distributions of co-authors are that,

Kyura illustrated the issues on the control in the servo parts of an industrial robot adopted in industry, numerical control working machine, three-dimensional measurement machine, a mechatronic machine called a chip mounter, etc.;

Nakamura explained these issues in systems control theory and formulized the obtained crucial points of problem solution;

Goto made computer simulations for the solution of these problems as well as verified the appropriation of these distinct theoretical results by using mechatronics-related experimental devices in the laboratory. In addition, among the undergraduate students, master students, doctor students who have interests in the control of mechatronic servo system, some items were allocated to them and the relevant achievements were obtained by research supervision. So far, about 60 conference presentations as well as 20 reviewed papers on the mechatronic control have been completed.

Based on the above research story, the motivation of writing this book was written down. Through the question answering in the conference for presenting the obtained research achievement or dealing with the paper reviewers or the conversation in the visiting the universities or research institutes which are doing research on robot manipulator, we felt strongly that a lot of researchers or engineers have many misunderstanding on the already solved problems in industry.

In fact, according to the words of coauthor Kyura, the strategies for the encountered problems in the servo controller design in industry depending on the experience with trial errors of designers and engineers are just responding to the demand of the world. These technologies have not become distinct in the so-called know-how world. Since they are not logical strategies, even successfully performing them, there are still many cases that the understandable explanation can not be obtained. In industry, even the clarification of the undesired points was conducted concretely, the contents are not announced. It is still in the present condition that why the good pursuit is hardly realized.

Through the collaboration, the essence of problems encountered in industry was analyzed and formulized logically and mathematically. According to the solution of derived equations and the verification of justifiability of these results, many useful items were obtained. At the present time, these items are summarized systematically. The opaque technologies under the name of know-how until now are explained distinctly. Therefore, many researchers or engineers can know them widely and effectively use them. These are the motivation of writing this book.

The problems discussed in this book are based on the common needs of industries rather than the pending problem areas of one research engineer in industry. The results for them, which were being caught empirically until now, are clarified logically. Therefore, the results are adapted for a real machine, and various performances or control methods of controller design previously determined with the experience of an expert can now are decided logically based on the adopted results. Moreover, a know-how only suitable for special situations until now, is changed into a more complicated and more ingenious universal technology. This book is unique in handling these problems.

The organization of this book is that, the design of the servo controller of mechatronic servo system is with respect to the fields of modeling, analysis and controller design control. It is from the introduction to the following chapters till 7.

In the introduction, the outline of mechatronic servo system and its main points of the problem in industry are given.

In chapter 2, these problems are solved reasonably, which are the achievements of cooperative research of co-authors. In each chapter, main points are attached.

The present conditions and problems in industry, main results, significant of results as well as the explanation of the main points of applications about each item are conducted at the commencement of each chapter.

It is acceptable even if the reader reads this book from the beginning. For the reader who wants to learn with the purpose of understanding, it is also good to learn each section of one chapter for dealing with the problems which are combined from the problems personally held and described in introduction. In each section of each chapter, main points are inserted at the beginning of each section for recommending the text reading thoroughly. The contents of each section are based on one of authors' papers which is specified with the quotation article number in the place of the bibliography list. Finally, the book contains an index, a glossary of terms, a collection of symbols and a description of the experimental devices used in our experiments.

During preparation, the book was read with distribution of sections of this book by seven master students of department of advanced systems control and engineering, graduate school of science and engineering, Saga university (Mr. Shigeto Aoki, Mr. Tatsuro Katafuchi, Mr. Daisuke Kushida, Mr. Kenta Shiramasa, Mr.Shojiro Yamagami, Mr. Masashi Tamura, Mr. Minoru Nishizawa). Referring to their impressions of the book, the book was revised to improve readability. The significance of the problems took up it in this book and the efforts are in making the essence of a problem to the formula appropriately. The keys to solution of many formulas are the easily adopted basis of classical control (Laplace transformation) or modern control (differential equation) learnt with the university bachelor degree, and the most fundamental knowledge in the control theory explained in appendix. Therefore, not only the enterprise directly related with system control or postgraduate students of university or researchers, but also the undergraduate students with the purpose to make the theory learnt in university into practice can be expected to read it widely. We expect that the knowledge obtained from this book can be adopted widely in mechatronic industries, and expect simultaneously that the research planted the root in this kind of ground will be expanded at the research institute etc. of an enterprise and, expecially and university.

At the end of the preface, since the materials of this book are all obtained from the cooperative research, the conditions of cooperative research, thoughts and feelings aroused from the cooperative research, are written as below, though it may be redundant.

1. The cooperative researcher should be proficient in each field.
2. Keep frequent discussion for a long time among cooperative researchers.
3. Respect the views of the partner mutually.
4. Fine mutual human relations.

Concerning the writing of this book, Mr. Kojiro Kobayashi, Department of production of the Morikita press, and Mr. Shoji Ishida, Department of compilation of the Morikita press, took care of it very much. All my great gratitude are here expressed.

October 1998   Editor-composer Masatoshi NAKAMURA

Contents

1 **Outline of Mechatronic Servo Systems** 1
  1.1 Emergence of Mechatronic Servo Systems 1
    1.1.1 Control Pattern of Mechatronic Servo Systems 1
    1.1.2 Characteristic of Servo System Applications 3
  1.2 Issues in Mechatronic Servo Systems 7
    1.2.1 Discussion on Modeling of a Mechatronic Servo System . 7
    1.2.2 Discussion on the Performance of One Axis in a
          Mechatronic Servo System 9
    1.2.3 Discussion on the Performance of a Multi-Axis
          Mechatronic Servo System 12
    1.2.4 Discussion on the Command of Mechatronic Servo
          Systems 14

2 **Mathematical Model Construction of a Mechatronic
  Servo System** 17
  2.1 4th Order Model of One Axis in a Mechatronic Servo System . 17
    2.1.1 Mechatronic Servo Systems 18
    2.1.2 Mathematical Model Derivation of a Mechatronic
          Servo System 20
    2.1.3 Determination Method of Servo Parameters Using a
          Mathematical Model 23
    2.1.4 Experiment Verification of the Mathematical Model ... 27
  2.2 Reduced Order Model of One Axis in a Mechatronic Servo
      System 29
    2.2.1 Necessary Conditions of the Reduced Order Model 29
    2.2.2 Structure Standard of Model 30
    2.2.3 Derivation of Low Speed 1st Order Model 31
    2.2.4 Derivation of the Middle Speed 2nd Order Model 32
    2.2.5 Evaluation of the Low Speed 1st Order Model and the
          Middle Speed 2nd Order Model 35

        2.3 Linear Model of the Working Coordinates of an Articulated
            Robot Arm 37
            2.3.1 A Working Linearized Model of an Articulated Robot
                  Arm 37
            2.3.2 Derivation of Adaptable Region of the Working
                  Linearized Model 42
            2.3.3 Adaptable Region of the Working Linearized Model
                  and Experiment Verification 51

3 **Discrete Time Interval of a Mechatronic Servo System** 53
    3.1 Sampling Time Interval 53
        3.1.1 Conditions Required in the Mechatronic Servo System . 54
        3.1.2 Relation between Control Properties and Sampling
              Frequency 56
        3.1.3 Sampling Frequency Required in the Sampling Control . 57
        3.1.4 Experimental Verification of the Sampling Frequency
              Determination Method 57
    3.2 Relation between Reference Input Time Interval and Velocity
        Fluctuation 58
        3.2.1 Mathematical Model of a Mechatronic Servo System
              Concerning Reference Input Time Interval 59
        3.2.2 Industrial Field Strategy of the Velocity Fluctuation
              Generated in Reference Input Time Interval 61
        3.2.3 Parameter Relation between the Steady-State Velocity
              Fluctuation and the Mechatronic Servo System 62
        3.2.4 Experimental Verification of the Steady-State Velocity
              Fluctuation 64
        3.2.5 Relation between Reference Input Time Interval and
              Transient Velocity Fluctuation 66
        3.2.6 Experimental Verification of the Transient Velocity
              Fluctuation 67
    3.3 Relationship between Reference Input Time Interval and
        Locus Irregularity 69
        3.3.1 Locus Irregularity in the Reference Input Time Interval 69
        3.3.2 Experimental Verification of the Locus Irregularity
              Generated in the Reference Input Time Interval 75
        3.3.3 Application Value of the Theoretical Analysis Result ... 77

4 **Quantization Error of a Mechatronic Servo System** 79
    4.1 Encoder Resolution 79
        4.1.1 Encoder Resolution of the Software Servo System 80
        4.1.2 A Mathematical Model and Resolution Judgement for
              Encoder Resolution 81
        4.1.3 Experimental Verification of the Encoder Resolution
              Determination 84

4.2		Torque Resolution 86
	4.2.1	Mathematical Model of the Mechatronic Servo System for Torque Resolution 86
	4.2.2	Deterioration of Positioning Precision Due to Torque Quantization Error 88
	4.2.3	Deterioration of Ramp Response Due to Torque Quantization Error 89
	4.2.4	Derivation of Torque Resolution Determination 93

5 Torque Saturation of a Mechatronic Servo System 97
5.1 Measurement Method for the Torque Saturation Property 97
  - 5.1.1 Torque Saturation of a Mechatronic Servo System 98
  - 5.1.2 Measurement of the Torque Saturation Curve and Experimental Verification 104
5.2 Contour Control Method with Avoidance of Torque Saturation 107
  - 5.2.1 Contour Control Performance with Torque Saturation and High-Precision Contour Control Method..........108
  - 5.2.2 Experimental Verification of Contour Control Considering Torque Saturation 114

6 The Modified Taught Data Method 121
6.1 Modified Taught Data Method Using a Mathematical Model ..121
  - 6.1.1 Derivation of the Modified Taught Data Method 122
  - 6.1.2 Properties Analysis of the Modified Taught Data Method 129
  - 6.1.3 Experimental Verification of the Modified Taught Data Method 133
6.2 Modified Taught Data Method Using a Gaussian Network 135
  - 6.2.1 Derivation of Modified Taught Data Method Using a Gaussian Network 137
  - 6.2.2 Experimental Verification for Modified Taught Data Method Using a Gaussian Network 142
6.3 A Modified Taught Data Method for a Flexible Mechanism ...144
  - 6.3.1 Derivation of Contour Control with Oscillation Restraint Using the Modified Taught Data Method 144
  - 6.3.2 Experimental Verification of Oscillation Restraint Control Using the Modified Taught Data Method......146

7 Master-Slave Synchronous Positioning Control..............149
7.1 The Master-Slave Synchronous Positioning Control Method ..149
  - 7.1.1 Necessity of Master-Slave Synchronous Positioning Control 150
  - 7.1.2 Derivation and Property Analysis of the Master-Slave Synchronous Positioning Control Method.............151

		7.1.3	Experimental Test of the Master-Slave Synchronous Positioning Control Method 153
	7.2	Contour Control with Master-Slave Synchronous Positioning .. 160	
		7.2.1	Derivation of the Contour Control Method with Master-Slave Synchronous Positioning................ 161
		7.2.2	Property Analysis and Evaluation of the Contour Control Method with Master-Slave Synchronous Positioning 163
		7.2.3	Experimental Test of the Contour Control Method of Master-Slave Synchronous Positioning................ 166

Glossary 169

Nomenclature 173

Experimental Equipments 179
    E.1 DEC-1 179
    E.2 Motoman... 180
    E.3 Performer MK3S 181
    E.4 XY Table 181

Appendix... 185
    A.1 Laplace Transform and Inverse Laplace Transform 185
    A.2 Transition Response 186
    A.3 Pole Assignment Regulator 187
    A.4 Minimal Order Observer 188

References.. 189

Index 193

1
Outline of Mechatronic Servo Systems

The mechatronic servo system is the major theme studied in this book. In particular, the servo system adopted in an electric servo motor is explained in this chapter. Several items of its utilization from the development stage to the present as well as its performances. The so-called mechanism machine (called as mechatronic servo system at the following), i.e., the servo system adopted in the numerical control machine or industrial robot, is generally different from the servo system introduced in the textbook of automatic control, which is very important when discussing the mechatronic servo system.

Firstly, the control pattern assigned in mechatronic servo system is illustrated. The properties of current servo system satisfying the control pattern and its utilization are introduced. Next, as the discussion items, the analysis on mechatronic servo system and its utilization are carried out.

1.1 Emergence of Mechatronic Servo Systems

1.1.1 Control Pattern of Mechatronic Servo Systems

The mechatronic servo system, as the control system satisfying the motion conditions of transfer axis of numerical control machine, was originally (about 1967) created when developing the DC servo motor. Then, in 1975 by Yaskawa Electric, the velocity control equipment (servo driver unit) unified the compensator of control system and power amplifier was sold. Initially, it was mainly adopted for the transfer axis control of working machine. From 1980, it was also adopted for the position and velocity controls of various kinds of mechanisms such as the industrial robot. At the generation of this mechatronic servo system, the control pattern, as the start point of servo system construction, is according to the following.

1. The velocity offset for step-shape torque disturbance is below n[rpm] (generally below 1[rpm]).

2. The velocity control ratio is one to several thousands (minimum 1[rpm] and maximum 3000–5000[rpm]).
3. The capability of power amplifier is effectively adopted (regulation time is shortened from the rated current acceleration/deceleration adopted for limited value).

Concerning the above three items, their necessity and significant in application are introduced respectively. In the transfer axis of a working machine, the pattern is determined from the motion of installed in tools for cutting or rotative cutting. There has the contact of the blades of these tools with processed product and the load to transfer axis as entering tools is the motion friction torque added constantly. When starting the process, there is negative force of processing in the transfer axis installed in tools. Certainly, the degree of negative force is different from the processing state. The negative force conducted at this time can be regarded as the step-shape torque load. This torque load is added as the torque disturbance to motor in control system. Therefore, at this time, the velocity offset is appeared, and the error of processed shape with designed shape is generated by the transfer axis. Hence, there exists the phenomenon of unexpected velocity offset due to torque disturbance.

The second item is necessary when a circular trajectory cannot be approximated by a polygon. In order to realize the circular trajectory, it is very difficult to accurately generate analogous sinusoidal or cosine instructions. Therefore, when generating the circular trajectory, the straight-line command approximating the circular trajectory by polygon is given with considering the velocity for the two-axis servo system constructing a plane. In order to move with constant speed along one edge of the polygon, two axes must move according to the velocity ratio corresponding to the axis incline. At the edges orthogonal between x axis and y axis, their velocity is infinite. To understand this case easily, the velocity command of driving system causing one axis motion should be needed from zero to infinite in theory. In fact, the edge number of approximated polygon is determined by the velocity control ratio of the driving system which can actually be implemented.

The third item is required by operational efficiency and power amplifier economics. The operational efficiency is evaluated by the actual operation time of the mechanism for an element, for example, the time of mechanism motion from beginning to end. Therefore, the time without cutting by knife is expected to be minimum. Moreover, a reduction in the time needed to reach the constant speed (regulation time of speed) is also attempted. However, it is not permitted to cost so much for this purpose. In general, the cost of a power amplifier is affected greatly by its output voltage and permitted maximal current. Thus, in the velocity control, it is required that the power amplifier is adopted with its maximal capability (allowance current) and the acceleration/deceleration time is shortened.

The structure of a mechatronic servo system designed for satisfying these performances is illustrated in Fig.1.1 for DC motor. As an aid to understanding the figure, generally, the position control is designed as ratio control and the velocity as well as current minor-loop in its inside is designed. Moreover, in the structure of power amplifier, PWM amplifier is always adopted. The carrier frequency of basic wave when using this PWM is from several to a few dozens [kHz] is used.

The structure component of this mechatronic servo system is changed from the original DC servo motor to an AC servo motor. Moreover, the controller using position, velocity, current loop can be also changed into a software servo system with a software algorithms using a micro processor, from the original hardware computing amplifier.

1.1.2 Characteristic of Servo System Applications

The emergence and structure of mechatronics have been briefly introduced in the former part. In order to understand that the usage of this mechatronic servo system is different from the general servo system, the main points are listed as below.

1. In a mechatronic servo system, there are two types of control. One is position control (PTP: point to point) emphasizing the arriving time and stop position from any position without considering the response route. Another is the contour control (contouring or CP: continuous path) emphasizing the motion trajectory from the current position to the next position (position at each moment and its motion velocity). These shapes are shown in Fig.1.2. The former one is the robot arm for element assembly, spot welding, etc, or used for the control of moving axis of mechanism for drilling a hole. The latter one is the arm of welding robot, painting robot, laser cutting robot, etc, or used for the control of transfer axis of mechanism implementing any three-dimensional shape processing (machine center, etc).
2. In the contour control, the servo system, as a position control system, requires strict velocity control for many kinds of response. Concerning the robot for welding, the importance of velocity control can be easily

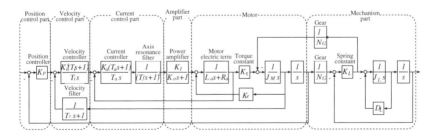

Fig. 1.1. Construction of position control system of one-axis mechatronic servo system

understood. In electric welding using an automatic welding machine, after setting voltage and current, the motion velocity of the torch (the tool spraying the fine solvent continuously after turning on the voltage) is determined according to the heat rate given along the curve of welding. Therefore, the motion velocity of this torch is changing while the given heat rate is also changing. If the over heat rate is thrown into, the mouth of relevant part is opened and the appropriate welding which should be with little given heat rate is impossible. In addition, for a painting robot, if the motion velocity of painting can is changed, the spot of painting is easily appeared. Besides, in the cutting operation of various materials, keeping the constant cutting velocity can guarantee the cutting quality.

3. In the contour control, an overshoot in the position control system should not occur. In many cases, velocity control system is also regulated so that the overshoot cannot occur. In the various kinds of actual processes, the generation of overshoot of position will cause fatal defect of shape. For example, in the process of constructing a shaft, if an overshoot occurs, the radius of the part becomes smaller, reducing the strength of this part. Moreover, if the vibrated trajectory exists insufficiency of shape, it cannot be revised at the later motion.

4. The objective command to servo system is obtained correctly before control in many cases. It can be said that, the element size, setting method, etc, of operation object of robot or process object of working machine can be completely known before starting the desired operation. In addition, the motion velocity at this time is also definitely determined. Therefore, the tract information necessary for motion is known before starting control. In addition, it can be supposed that external disturbance is mixed into the control system. When the mixed disturbance over the supposition, concerning the safety of equipment, the motion of control system should be stopped and the power source for driving should be isolated

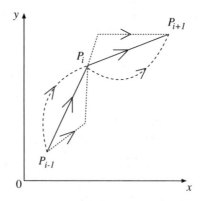

Fig. 1.2. PTP control and CP control

as far as possible. That is to say, the motor must be selected from the clearly discussed results on the necessary maximal torque for executing operation. In addition, the size of continuously mixed disturbance must be below the continuous rated torque of motor.

5. In many servo systems, a feedback system can be only established based on the information of servo actuator, but not according to the information of each moveable tip or motion tip. It means that, the detector of position and velocity in the opposite load side of motor (side without load) is installed and then the feedback system of actuator control is resembled by the obtained information. This kind of control system is called a semi-closed loop. Generally, it is very difficult to construct the feedback system by motion tip information in many mechanism machines. The structure of a full-closed loop on the feedback of moveable tip information adopted in some parts is shown in Fig.1.3. In addition, almost all mechanic structures of industrial six-freedom degree robots are semi-closed loop. The relation with servo actuator is briefly shown in Fig.1.4. The structure of this kind of semi-closed loop cannot be obtained in the mechatronic servo system as same as the general feedback system. For taking into account the system as same as the general feedback system, the condition is that the system should be rigidly unified with the actuator when mechanism is within the control region according to the desired motion command.

6. The actuator installed in the mechanism structured for multi-axis moveable mechanism generally corresponds to the forward motion of one actuator as well as rotation of one axis (freedom degree). The arbitrary curve in three-dimensional space implemented by simultaneous control of multiple axes is given in a servo system as the command of time function about the position for desired motion in each independent axis. The precondition in control system is that axis is regarded as independence. In fact, for example, in the case of a multi-axis robot arm, the reaction of one axis motion affects other axes, i.e., axis interference occurs. This axis interference is very important when trying to minimize it in mechanism design. Moreover, in a mechatronic servo system, when considering one axis, the effect from other axes due to its reaction is regarded as the disturbance.

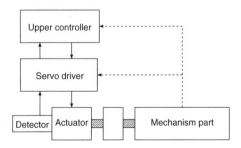

Fig. 1.3. Structure of industrial servo system

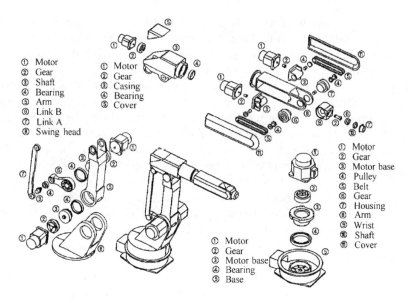

Fig. 1.4. Structure of industrial robot and arrangement of actuator

And for reducing the effect in control to a minimum, the motion of axis should be changed as be capable of independence.

7. The generation of objective reference input for realizing mechanism movement structured with a multi-axis mechanism is by the servo system in which the independent motion of each axis including introduced machine mechanism can be realized. The features of reference input is regulated for keeping the consistence of each axis. In almost all cases, position control system is regarded as a 1st order system. The feedforward gain should be identical. If regulating like this, it is very easy to implement the algorithm of reference input generation to a multi-axis servo system for any spatial curve.

8. For realizing an arbitrary curve in three-dimensional space, in most of cases, the curve is approximated by a folded line. As its results, the reference input to each axis servo system is renewed in each given coordinate point and the ramp input with various slope angles is given continuously. The velocity of each axis is calculated for making the given synthetic velocity as a desired velocity. In addition, in the case of performing the acceleration/deceleration control at start/stop point, the reference input for simultaneous start/stop of all axes, i.e., same command to each axis at acceleration/deceleration point, is generated.

9. The data regarded as objective reference input to servo system from the upper control device, such as computer or special control, should be given according to the designated period (designated time interval). Therefore, the reference input for servo system is described with the form of a velocity

command. Here, the period of this command or time interval is called as the system clock of the servo system in the controller or servo data rate of the controller. Since this reference time interval is selected based on the property of the servo system dependent on the mechanism part structure and related with the capability of control devise, its value represents the synthetic performance of a machine. In the numerical control device of a working machine, several [ms] as well as ten and several or several ten times [ms] are adopted.

The knowledge of industrial expects cannot be understood definitely. Therefore, so far, the theoretical analysis on the properties of the control system structure, as well as the various properties of mechatronic servo system taking into account its utilization, as above, cannot be found. For this purpose, in this book, stepping on the utilization of mechatronic servo system, various adopted control methods and realized control performances by these methods are firstly discussed theoretically or arguably as the main point, and then the discussion on the development in the future is added.

1.2 Issues in Mechatronic Servo Systems

In order to understand the current mechatronic servo system and develop servo systems with better performance than at present, this servo system must be investigated from various points of view. The discussion points are listed as below.

1. Modeling of a mechatronic servo system
2. Performance of one axis in a mechatronic servo system
3. Performance of multiple axes in a mechatronic servo system
4. Command to servo system

The above viewpoints come from the system components of servo system in theory. It means that a servo system is one of the system components for establishing a mechanism machine and needed to know that in which step servo system can be thought as good enough so that the system is constructed efficiently, desired mechanism as well as performances is realized, etc. Therefore, the description sequence of subsequent chapters is different from the explanation sequence in this chapter. Each section is divided by the items listed above.

1.2.1 Discussion on Modeling of a Mechatronic Servo System

From the view of using the model of mechatronic servo system, this model should be divided into two points. One is the model with the servo system not only taking into account the mechanism structure but also the load. Another is the model combining the modeled mechanism structure and servo system.

For the mechanism performing orthogonal motion, various discussions can be carried out only by the former modeling. But for the machine as an articulated robot, the latter modeling is also necessary.

(1) Modeling of the Overall Mechatronic Servo System

In the mechatronic servo system adopted in any mechanism, such as a numerical control working machine, industrial robot, etc., representing the industrial mechanism machine, the property of previous or present servo system can be expressed by K_p. In general, the value of K_p is the high rigid of huge machine. In the general rotation plate, machine center, etc, the value of K_p is 35–40[1/s]. In an industrial robot, the value of K_p is 15[1/s]. It is naturally the most simple approximation 1st order system in control system. However, concerning the mechatronic servo system, it should be considered which conditions must be satisfied in its internal structure, and additionally, it is nearly not clear about the usage condition of this 1st order delay system. Actually, if the maximal speed used in this machine is about 1/10 speed region, this approximation can model the whole system quite well. However, if it is number for one speed, it will have big deviation with the actual system.

In analyzing the current servo system, the mechanism is thought to combine with shaft of motor which is as rigid. Under this precondition, the servo motor for a driving mechanism is selected. The control parameter regulation of such a servo system is also following this consideration. Therefore, the 1st order approximation with precondition is pursued to be clarified[4]. However, in fact, it is difficult to satisfy this precondition due to various restraints. High-speed and high-precision motion of mechatronics machines has been the objective in recent years. For finding out the control strategy, it is required to model control system correctly.

Concerning about this problem, it is explained in 2.1 and 2.2.

(2) Modeling of a Multi-Jointed Robot

Generally, in the multi-jointed industrial robot, orthogonal motion (in working coordinate) data is generated by using coordinate calculation based on inverse kinematics. By the servo system for joint angle control (in joint coordinate), the motion can be realized in working coordinate system as the orthogonal coordinate system. The inverse kinematics calculation of orthogonal coordinate value is performed at each reference input time interval.

When given two points for performing orthogonal motion, with high-speed motion, the phenomenon of deviation of several millimeters in the motion trajectory of the line between two points is apparent. The motion velocity of this time is about 1[m/s]. The reference input time interval of robot controller is generally adopted with about 20[ms]. This velocity is lower than the velocity of appearing centrifugal force rated with two times velocity for general issue or collision rated with two axes velocity integral. Therefore, the trajectory deviation is difficult to consider based on these effects. But if lengthening the reference input time interval or whether or not orthogonal motion happening, several reasons should be considered.

When the velocity of present contour control is below 25[m/min], the trajectory deviation does not occur. Therefore, in the control based on the previous position decision control concept, the trajectory precision can be required. In the position decision control, the motion with the highest velocity allowed by this robot can be performed in almost all cases. In the actual examples of these kinds of application, such as hard-cutting, spot-welding, etc, the position variation (trajectory precision) is several [mm]. Recently, the following is also required.

In order to analyze the control strategy for satisfying these requirements, the correct modeling for multi-jointed robot is needed. The relevant detail description is given in section 2.3. The discussed modeling combining the modeling of the whole servo system in the former part, the importance of modeling control system in future mechanism machines is illustrated.

1.2.2 Discussion on the Performance of One Axis in a Mechatronic Servo System

In a usual, mechatronic servo system consists of multi-axis mechanism. When taking into account the performance of a mechanism machine, the analysis on multi-axis servo system must be carried out. However, the structure for this actuator is basically independent for one axis. For the basic feature of a mechatronic servo system, the discussion based on the state of one axis structure is sufficient.

Hence, there are two problems on discrete time interval when analyzing the one axis performance of mechatronic servo system. One is that the structure of current mechatronic servo systems are almost all software servo systems and they must be thought of coming from the sampling control systems. Therefore, the data renewal time interval of control system is determined by sampling frequency. In general mechatronic servo system, there exist same delay time and 0th order hold with this time interval. Therefore, this time interval greatly affects the characteristic of closed-loop control system.

Another is that, the upper controller seen from the servo system, i.e., the computer using for internal trajectory calculation of the controller, is performed in a time interval providing command given in the servo system. From the relation between this time interval and performance of the control system, the overall mechatronic servo system performance of a mechatronic machine can be determined. From this point of view, the value of these discrete time intervals are very important for analyzing the performance of a mechatronic servo system.

(1) Proper Sampling Frequency

In the middle of 1980s, microprocessor (CPU), i.e., digital signal processor (DSP) became cheaper. These processors are equipped into closed-loop of servo system. Hence, servo system is constructed and movement can be remarkably flexible. Software servo systems were developed.

These servo systems were developed in the laboratory belonging to one of authors. From the experiments, an experience rule, was obtained. The eigenvalue of position control system using for a mechatronic machine based on the realized software servo from the analogue velocity of a control device cannot be over about 1/30 sampling frequency. Moreover, the velocity control system is made by the software servo system and its inside can be found similar with the analogue pattern.

The great difference here between the general sampling control system and the control system used in the mechanism machine is the delay time. In the usually equipped process control, comparing with the sampling time interval, the consumed time for working out the state input and operation value can be neglected. However, in the servo system of a mechanism machine, this cannot be successful.

If the software servo system adopted in a mechanism machine is the object of simulation, various unknown parts are closed up. How to set the property of power amplifier with PWM pattern, and how to catch the timing of state input and the dynamic of operation output can be obtained.

In general, in a software servo system, a very big sampling frequency is adopted. Namely, under the restraint of hardware cost, the maximal sampling frequency is selected. In determining the sampling frequency by this way, the performance boundary of the servo system when using this frequency is not distinct. Even though expecting to raise its performance, which component of control system should be improved is also unknown.

In section 3.1, the quite simple form of mechatronic servo system was analyzed. The relation between the performance of a control system and sampling time interval when considering the utilization situation of a servo system was clarified.

(2) Reference Input Time Interval

When considering the characteristic of a mechatronic servo system as introduced before, and regarding the loop structure of a control system about actuator above investigation of servo system characteristic as the identical important item with its controller design, how to provide the command to servo system is a problem. This problem is about the form of time function of command. The problem of command containing the way of data given must be discussed.

In the discussion of this command system, with the current controller structure, as the item about the control performance of a servo system, the time interval of given data to the servo system through the interface from the upper controller is expected. Generally, in the controller of the mechanism machine, the data to the servo system is given in a designated period. This designated period is called the reference input time interval. This is also called the (controller) system clock.

This reference input time interval is discrete width as the data to the servo system. Within this interval, the command function of each axis is calculated. Then, this calculated value is obtained in the servo system with the state of zero order hold. From this, the motion of the servo system generates velocity periodic variation relied on this time interval and trajectory deterioration.

Previously, the reference input time interval is obtained as the value representing the controller performance. At present, in the newly developed controller, this value has the trend to be minimal. However, dominated by the development of the microprocessor, the desired performance is expected to be realized without great cost. Therefore, the reasonable explanation of the relationship between this reference input time interval and various generated phenomenon is almost non-existent.

The competition of mechatronic product cost is rapidly increased. High performance is required meanwhile keeping the current situation. In this situation, the performance of servo driver unit, the performance of the upper controller (reference input time interval, etc) as well as the characteristic of load are analyzed comprehensively. By taking these performances obtained the balance when observing these performances respectively as the whole, it is very important to realize these desired performances comprehensively. As the first stage for analyzing them, from the view of the servo system characteristic of one axis, the discussion on the reference input time interval is carried out in section 3.2 and 3.3.

(3) Quantization Error and Control Performance of Control System

The structure of the software servo system was developed from only the position controller software to both velocity controller and current controller software, from the development of utilized CPS, i.e., DSP. In the construction of the control system, high response performance is generally required from its internal minor-loop. In the electric servo system, current feedback loop is the inner-most loop. How resolution of current detection is expected for satisfying the required performance of servo system is an important item to discuss in the stage of designing hardware constructing servo system.

As usual, although the control performances about the position and velocity of the servo system were clear, the theoretical equations for expressing the design which the control performance must be satisfied about its internal is unknown. In view of the concrete circuit structure, the discussion of the item on quantization error is formulated. However, the analysis solution on various internal parameters relation to the control system structure is very difficult to solve. Its difficulty would be estimated by taking into account the equation expression of power amplifier of PWM pattern. From this point of view, in almost all present cases, the quantization scale of control system internal, i.e., resolution is determined based on experience.

Here, for the current (torque) loop of the motor, the most internal loop of mechatronic servo system, the relationship between the necessary performance in the control system and the resolution of current detection part is investigated. In order to clarify the main issue on considering the current structure of the mechatronic servo system, the foreseen whole control system is considered and the problems are formulated.

On these problems, is discussed in chapter 4, after analyzing the resolution of position detection firstly in section 4.1, the torque resolution is investigated in section 4.2. From the formulation illustrated here, the resolution of torque command considering velocity variation ratio as a control performance is clear. According to this result, the necessity of identical precision with the necessary resolution in current detection is clear. Moreover, in the case of zero-zone of power amplifier, i.e., nonlinear characteristic, it is easy to evaluate that the high resolution is necessary from the obtained results here.

1.2.3 Discussion on the Performance of a Multi-Axis Mechatronic Servo System

The basic part of on discussion on mechatronic servo systems can be carried out as a one-axis servo system. However, when investigating the performance of mechanism machines, they must be investigated as multiple axes. The motion of multi-axis servo systems causing basic phenomena due to torque saturation can be found. When using a servo system in the state of one axis, there is almost no problem in the induced phenomenon due to torque saturation from the servo system performance point of view. However, if this phenomenon occurs in the multi-axis contour control, it will produce great effects on servo system performance. These problems are discussed in section 5.1 and 5.2.

(1) Torque Saturation

Generally, in mechatronic servo systems, the ratio of maximal torque that can be used in rated torque and acceleration/deceleration is about 1:3~5. In actual servo systems, constant coulomb friction from motion resistance occupies a big part of rated torque when the servo system is set into the mechanism. It means that, the opposite force in operation is regarded as the torque load. In order to allow these torques in the control system when performing the movement along a straight line, their values are reduced remarkably. Hence, in contour control, the servo system must guarantee the movement along the straight line. When clarifying the application condition of the mechatronic system, it must grasp that in which scale torque reaches saturation in the state of capable motion of the mechanism as well as in which degree control performance deteriorates due to torque saturation.

The mechatronic servo system design should select a servo motor for the driving mechanism in many cases except the stage of research. Therefore, in servo motor selection, the velocity profile (stage form) for driving is designed and acceleration/deceleration as well as constant motion torque for designated parameters (acceleration time, maximal velocity, etc) are calculated. In the mechatronic servo system driven by the motor selected as above, it is almost impossible to consider clearly the torque reflecting the actual adopted status.

In section 5.1, firstly, the measurement method of torque saturation is shown. Based on this method, the torque saturation of the actual mechanism with the different statuses of a single motor can be known. Moreover, from grasping the occurred phenomenon when existing torque saturation as in the above illustration, the reason of actual phenomena can be definitely judged. For avoiding torque saturation naturally, the actuator capable of exporting torque with big capacity is needed to use. In reality, the correct motion is more important than changing the application method. For this purpose, it is necessary to know the simple avoidance method, which is discussed in section 5.2.

(2) Master-Slave Synchronous Positioning Control

The master-slave synchronous positioning control method is the control that must satisfy the ratio relation between the movement of one axis and that of another axis between two axes. This control is generated from the motion performance required in the tapping process of the machining center. In the tapping process, the 1st axis is the master axis of the machining center. This axis is moving as the control system performing start and stop for stage form driving installed in the rotational tools. The transfer axis as the second axis should be traced, namely synchronized. So then it performs a parallel movement as a position control part. When the tapper, a tool for standing tap for rotation movement kept in the master axis, is rotated once, the transfer axis must be moved correct one pitch of the spring. Since this correct motion cannot be guaranteed, the tool called a soft-tapper is used to keep the tapper through the spring and the synchronous error of rotation and transfer is absorbed.

However, since this soft-tapper is very expensive, for the decrease of running cost for tapping, high-precision master-slave synchronous positioning control is demanded. In the middle of the 1980s, not only was the present soft-tapper adopted, but also the general tapping process was realized. At this time, the rotation times of master axis is from 3000 to 4000[rpm].

In order to improve production in the future, high-speed master axis rotation number is demanded. For the relevant high-level spring, high-precision master-slave synchronous positioning control is required as well. The relevant discussion is carried out in section 7.1. In addition, the possibility of adapting this master-slave synchronous positioning control in contour control is explained in section 7.2.

1.2.4 Discussion on the Command of Mechatronic Servo Systems

For improving the motion performance of the whole mechatronic servo system, the method for providing the command to the servo system at each moment is a very important factor. It means that the final desired motion of current mechatronic servo systems should be approximated from the known information before the beginning of control. As the precondition of use state of present servo system, the revision method for the known command for realizing the desired motion is analyzed in chapter 6.

(1) Modified Taught Data Method

The contour control for a three-dimensional curve in the present industrial robot, the curve is approximated by a folded line. In the contour control, the locus (position) as the form and its motion speed are the important control parameters. As usual, the ramp input with a designated slope for each axis as its command is introduced.

In such a kind of robot performing this control, when given three points and angles are described by a line trajectory, at the corner part, the trajectory deviates from the corner point depending on the velocity. Certainly, the velocity is also decreased. For dealing with this kind of situation, skilled operator of teaching is successfully carried out by given taught data varied from the final needed shape for eliminating the deviation from the corner point in continuous motion. This method is illustrated concretely for realizing desired motion by revising commands to servo system.

When approximating a robot arm by a 1st order delay system and assuming it as an orthogonal coordinate robot, concerning the quite long straight line, the theoretical explanation of this phenomenon realizing skilled operator can be easily carried out. However, it is known that this method is almost impossible by the above formulation when it is adopted for the general multi-axis mechatronic servo system.

In sections 6.1, 6.2 and 6.3, the solution of the method for improving the effective motion performance by the taught point which is a circular trajectory but not linear trajectory, namely the composed trajectory is given. Here, taking the taught point information as the desired final robot motion, the analysis solution for the issue, that how the taught point information is revised to be given in a servo system so that the desired motion can be desired, is illustrated. The flowchart is expected to remember the solution whose roots must be definitely used in a mechatronic servo system.

The command method adopted in this book is not only introduced in this chapter, but also considered for the performance improvement of the future mechatronic servo system. The command method to the servo system, considering the properties of mechanism, the features of disturbance size, the features of process conditions, etc., is regarded as an important item similar with the servo system characteristic improvement as the feedback loop.

2

Mathematical Model Construction of a Mechatronic Servo System

In this chapter, from the view of servo controller design of mechatronic equipment, such as an industrial robot, NC machine tool, chip mounter, etc, and stepping on the action of a mechatronic servo system driven by the signals of a power amplifier, the4th order model expressing faithfully the action observed from the appearance, the reduced order model simplifying the 4th order model according to the action condition and the approximated linear 1st order model in working coordinate are introduced.

These models are constructed for the characteristic analysis of mechatronic servo system and the design of servo controller. The mechatronic systems model introduced in this chapter are the basis of all analysis and design in the following chapters. Each model is the general linear model in terms of the form. In the deduction of this equation, the characteristics of the actual mechatronic servo system can be expressed correctly with this simple equation for the first time, according to adopting appropriately the actual restriction conditions of mechatronic servo systems of the industrial field.

2.1 4th Order Model of One Axis in a Mechatronic Servo System

In the determination of parameters of a mechatronic servo system controller, such as position loop gain K_p, velocity loop gain K_v, etc, as well as in the discussion of control strategies adopted in the controller; it is necessary to construct a mathematical model expressing the action characteristic of mechatronic servo system appropriately. In an industrial field, determination of parameters of the servo controller is mostly based on the empirical rule of practician. There is no mathematical model comprehensively expressing the mechatronic servo system including all mechanism parts, servo motor, servo controller, etc.

Since the structure of mechatronic servo systems in industry is the high order for expressing all factors, the 4th order model as (2.8), which retains the necessary parts taking into account the servo properties in the general mechanism by eliminating the unnecessary properties of the servo amplifier converter inverter, etc, from the view of servo controller design, is proposed.

This 4th order model correctly expresses the response characteristic of one axis of a mechatronic servo system. In the mechatronic servo system with a multi-axis structure, this 4th order model can be expressed by combining several independent axes. For realizing the expected action characteristics of a mechatronic servo system, the relation between the necessary servo parameters (K_p, K_v) and natural angular frequency (ω_L) of mechanism part, called as empirical rule, is $K_p \leq K_v/6$. The appropriation of this equivalent relation ($c_p = 0.24, c_v = 0.82$) can be theoretically shown in the 4th order model. In addition, by using this mathematical model, the various control properties of the mechatronic servo system can be analyzed and they can be adopted in the design of the servo controller.

2.1.1 Mechatronic Servo Systems

(1) Structure of an Industrial Mechatronic Servo System

Fig. 2.1 illustrates the whole structure of a mechatronic servo system. As shown in this figure, the industrial mechatronic servo system is the servo system including the **mechanism part**, the servo motor driving **axis** included in the mechanism part, the servo motor and the **servo controller**. In this system, the **management part** managing the entire mechanism part and the **reference input generator** are separated. The servo system of each axis is constructed by the **motor part**, the **power amplifier part**, the **current control part**, the **velocity control part** and the **position control part** and sensor (position detector, velocity detector, current detector) in order to detect the signal from various parts, and connected with the mechanism part by hardware.

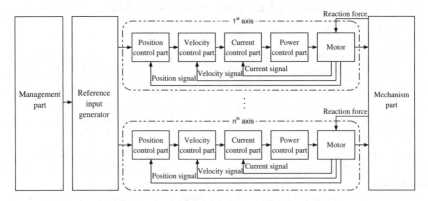

Fig. 2.1. Industrial mechatronic servo system structure

2.1 4th Order Model of One Axis in a Mechatronic Servo System

In an industrial mechatronic servo system, the servo system of each axis is always controlled independently (refer to 1.1.2 item 6,7). Actually, the interference or friction of each axis is different according to the structure of the mechanism part. Although it is possible to design an optimal servo controller corresponding to the various mechanism, the cost of designing a servo controller respectively for each mechanism became very high and hence the implementation in industry is very difficult. Therefore, for designing a servo system which can be adopted, this servo system is combined into each axis corresponding to the mechanism part. That is to say, in an industrial mechatronic servo system, the discussion on servo system is carried out for only one axis, because of the importance of the servo problem for each axis. If this one axis problem can be solved, the general industrial mechatronic servo system problem can be solved, when each axis of an industrial mechatronic servo system is simply combined and the characteristic in joint coordinate can be analyzed approximately by characteristic in working coordinate by using nonlinear coordinate transform between the working coordinate and the joint coordinate about articulated robot arm.

(2) Servo Controller of Industrial Mechatronic Servo System

The block diagram of servo system of each axis in a mechatronic servo system is a 13th order or higher high-order system strictly illustrated in Fig. 1.1(refer to item 1.1.2). From Fig. 1.1, the information of locus is not feedback in the servo controller. From this 13th order model, the features of modeling from the point of the servo controller of a mechatronic servo system is summarized as [4]:

1. The power amplifier can be obtained linearly when a big carrier frequency is designed greatly;
2. The dead zone of the power amplifier can be neglected;
3. The resonance frequency of axis torque of each axis motor is about 5~8 times that of the natural frequency of the mechanism part and can be neglected when eliminating axis resonance by an axis resonance filter;
4. The cut-off frequency of the velocity detection filter and axis resonance filter can be neglected if it is higher than the natural frequency of the whole mechatronic servo system;
5. The current control part is designed by considering the balance of the electric features of motor;
6. The position detection is obtained by the logical calculation of two pulse signals of the encoder and judgement of direction and increase/decrease. The countering of the pulse without noise in the pulse counter is considered;
7. The delay in response can be neglected if the response velocity of velocity detection is higher than the response velocity of the mechanism;
8. The torque disturbance is compensated in the integral (I) action of PI controller of velocity loop.

According to these characteristics, the original complex structure of industrial mechatronic servo systems can be simplified using a simple mathematical model in the contour control.

2.1.2 Mathematical Model Derivation of a Mechatronic Servo System

(1) 4th Order Model of an Industrial Mechatronic Servo System

For combining the mechanism part of a mechatronic servo system and the mechanical part of the motor, a two mass model is adopted[5, 6]. The two-mass model is the model in which the inertial moment of the motor and the inertial moment of the load are connected by a spring. The motion equation in the motor side and the mechanism part side can be written as below, which including the inertial moment of motor J_M, the rotation angle θ_M, the inertial moment J_L of load in the mechanism part, the viscous friction coefficient D_L, the whole spring constant K_L with the gear for connecting the mechanism part and motor axis, gear ratio N_G and torque T_M generated in the motor side,

$$J_M \frac{d^2\theta_M(t)}{dt^2} = T_M(t) - T_L(t) \tag{2.1}$$

$$J_L \frac{d^2\theta_L(t)}{dt^2} = N_G T_L(t) - D_L \frac{d\theta_L(t)}{dt} \tag{2.2}$$

$$T_L(t) = \frac{K_L \left(\theta_M(t) - N_G \theta_L(t)\right)}{N_G^2} \tag{2.3}$$

where $T_L(t)$ in (2.3) is the reaction force added on the motor side from the mechanism part side. However, the friction of the motor itself is ignored because it is too small. When equations (2.1) and (2.2) are transformed by a Laplace transform (refer to appendix A.1), the transfer function of the two mass model is as

$$\theta_M(s) = \frac{T_M(s) - T_L(s)}{J_M s^2} \tag{2.4}$$

$$\theta_L(s) = \frac{N_G}{J_L s^2 + D_L s} T_L(s) \tag{2.5}$$

$$T_L(s) = \frac{K_L}{N_G^2} (\theta_M(s) - N_G \theta_L(s)). \tag{2.6}$$

Concerning the servo controller side, from the features 3, 4, 5, 8 of an industrial mechatronic servo system introduced in 2.1.1(2), the influence of the axis resonance filter feature and velocity detection filter feature in Fig. 1.1 can be neglected due to their slightness. When making the current loop transfer function in the servo controller as one and the velocity controller is expressed as P control, the transfer function of the servo controller and the electric part of motor is changed as

2.1 4th Order Model of One Axis in a Mechatronic Servo System

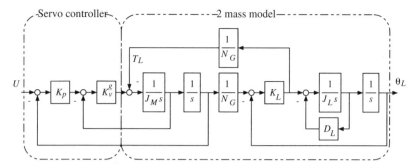

Fig. 2.2. Block diagram of 4th order model of industrial mechatronic servo system

$$T_M(s) = K_v^g[K_p\{U(s) - \theta_M(s)\} - s\theta_M(s)] \quad (2.7)$$

where $T_M(s)$ in equation (2.7) denotes the torque generated from the motor. The first item of (2.7) is the transfer function of the servo controller. The second item expresses the influence of the reaction force $T_L(s)$. $U(s)$ is the angle input to the motor. K_p is position loop gain. K_v^g is velocity amplifier gain.

The transfer function from the angle input $U(s)$ for the motor of the whole mechatronic servo system to the angle output $\theta_L(s)$ of the load can be written as below, when deriving the relation equation between $U(s)$ and $\theta_L(s)$ by eliminating $\theta_M(s)$, $T_L(s)$, $T_M(s)$ from four relation equation (2.4)~(2.7) with five variables $U(s)$, $\theta_L(s)$, $\theta_M(s)$, $T_L(s)$, $T_M(s)$ (refer to Fig. 2.2).

$$G(s) = \frac{a_0}{N_G(s^4 + a_3 s^3 + a_2 s^2 + a_1 s + a_0)} \quad (2.8)$$

$$a_0 = \frac{K_L K_p K_v^g}{J_L J_M}$$

$$a_1 = \frac{K_L K_v^g}{J_L J_M} + \frac{D_L K_p K_v^g}{J_L J_M} + \frac{D_L K_L}{N_G^2 J_L J_M}$$

$$a_2 = \frac{K_L}{J_L} + \frac{D_L K_v^g}{J_L J_M} + \frac{K_p K_v^g}{J_M} + \frac{K_L}{N_G^2 J_M}$$

$$a_3 = \frac{D_L}{J_L} + \frac{K_v^g}{J_M}.$$

This 4th order model of a mechatronic servo system can be effectively adopted in the development of servo parameter determination or control strategy.

In the actual mechatronic servo system, for changing velocity controller as PI controller, it is as shown strictly in the block diagram of Fig. 1.1. To this controller, in the 4th order model of Fig. 2.2, velocity controller is expressed by an equivalent P control. The integral (I) action in velocity controller in the actual mechatronic servo system is performed for torque disturbance compensation. The time shift of output response is nominated by the gain of P control. On above way, the ratio gain K_v^s of PI control in the general motion of an actual system is not the velocity amplifier gain K_v^g in the model of Fig. 2.2, but is expressed by the ratio gain when PI Controller is equivalent to the P control.

(2) Normalized 4th Order Model for Servo Parameter Determination

The parameters of the servo controller in the 4th order mode (2.8) are position loop gain K_p and velocity amplifier gain K_v^g. Concerning the velocity amplifier gain K_v^g, the total inertial moment transformed from the motor axis with a rigid connection is assumed as

$$J_T = J_M + \frac{J_L}{N_G^2}. \tag{2.9}$$

K_v is defined as the velocity loop gain by using this J_T as

$$K_v = \frac{K_v^g}{J_T}. \tag{2.10}$$

This velocity loop gain is regarded as a servo parameter. Hence, position loop gain K_p and velocity loop gain K_v has the same order for using later. In addition, in equation (2.8), by viscous friction coefficient D_L, spring constant K_L and load moment of inertia J_L, the natural angular frequency ω_L and damping factor ζ_L expressed by the features of mechanism part is written as

$$\omega_L = \sqrt{\frac{K_L}{J_L}} \tag{2.11a}$$

$$\zeta_L = \frac{D_L}{2\sqrt{J_L K_L}}. \tag{2.11b}$$

When expressing the general features of the mechanism part, for convenient expression by natural angular frequency ω_L and damping factor ζ_L with viscous friction coefficient D_L and spring constant K_L, ω_L and ζ_L are adopted as the parameters of the mechanism part.

The 4th order model derived in the last part is determined by the natural angular frequency ω_L and damping factor ζ_L as the features of the mechanism part, as well as the servo parameter K_p, K_v. However, since the natural angular frequency of the mechanism part has a strong dependence on its size or mass, it is expected that the standard determination of servo parameters is not based on the natural angular frequency of the mechanism part. Therefore, the position loop gain K_p and velocity loop gain K_v are expressed as below by using the natural angular frequency ω_L of the mechanism part as

$$K_p = c_p \omega_L \tag{2.12a}$$

$$K_v = c_v \omega_L. \tag{2.12b}$$

2.1 4th Order Model of One Axis in a Mechatronic Servo System

It is the transformation of equation (2.8) using c_p, c_v in equation (2.12a) and (2.12b). When we put equation (2.11b)~(2.12b) into (2.8), the normalized 4th order model without dependence on natural angular frequency ω_L is derived

$$G_c(s) = \frac{b_0}{N_G(s^4 + b_3 s^3 + b_2 s^2 + b_1 s + b_0)} \tag{2.13}$$

$$b_0 = (1 + N_L) c_p c_v$$
$$b_1 = (1 + N_L)(c_v + 2 c_p c_v \zeta_L) + 2 N_L \zeta_L$$
$$b_2 = (1 + N_L)(1 + 2 c_v \zeta_L + c_p c_v)$$
$$b_3 = 2\zeta_L + (1 + N_L) c_v$$

where

$$N_L = \frac{J_L}{N_G^2 J_M} \tag{2.14}$$

is the ratio between the inertial moment and motor axis equivalent inertial moment of the mechanism part. By using this normalized 4th order model (2.13), the common discussion on the arbitrary natural angular frequency ω_L of the mechanism part can be carried out.

2.1.3 Determination Method of Servo Parameters Using a Mathematical Model

(1) Control Performance Required in an Industrial Mechatronic Servo System

The response characteristic of an industrial mechatronic servo system is required to have a fast response in the system within the region where there is no generation of oscillation and overshoot (refer to 1.1.2 item 3). Previously, the servo parameters are determined by satisfying the requirement based on the test error or experience. The proper determination method can be derived by a normalized 4th order model (2.15) here

In an industrial mechatronic servo system, the following conditions are successful:

- The motor is selected when the moment of inertia J_M of the motor is satisfying $3 \leq N_L \leq 10$ from the moment of inertia J_L of the mechanism part and gear ratio;
- The damping factor ζ_L of mechanism part is $0 \leq \zeta_L \leq 0.02$.

For the latter condition, since the damping factor ζ_L is very small in an industrial mechatronic servo system, then $\zeta_L = 0$. However, $\zeta_L = 0$ is existed in the situation of continuous oscillation generation which is the most difficult to control. Then this assumption is sufficient for this situation. When put $\zeta_L = 0$ into equation (2.13), it can be as

$$G_c(s) \approx \cfrac{1}{N_G\left\{\cfrac{s^4}{(1+N_L)c_p c_v} + \cfrac{s^3}{c_p} + \cfrac{(1+c_p c_v)s^2}{c_p c_v} + \cfrac{s}{c_p} + 1\right\}}. \qquad (2.15)$$

From the current utilization of an industrial mechatronic servo system, there are the following conditions for servo parameters determination satisfying the desired control performances

1. There are two real poles and one complex conjugate root in the normalized 4th order model (2.15) (condition A)
2. The response component of the complex conjugate root is smaller than the response component of the principal root (condition B).
3. The response component of the complex conjugate root is more quickly converged than the response component of the principal root (condition C).
4. If satisfying the above three conditions, the servo parameters K_p, K_v can be determined for a faster response.

(2) Ramp Response of the Normalized 4th Order Model

For determining the servo parameters satisfying the required control performance introduced in 2.1.3(1), the ramp response of the normalized 4th order model (2.15) should be worked out. The reason for using a ramp response is that, the ramp input can be adopted in each axis of an industrial mechatronic servo system in almost all contour control (refer to 1.1.2 item 8).

For the ramp response of the normalized 4th order model, ramp input is $u(t) = vt$. From condition A, there are given two poles as $-\tau_1, -\tau_2$ ($\tau_1 < \tau_2$) and one complex conjugate root $-\sigma + j\rho, -\sigma - j\rho$, and the ramp response is calculated as (refer to appendix A.2)

$$\begin{aligned} y_4(t) = & \left(t - K_0 + K_1 e^{-\tau_1 t} + K_2 e^{-\tau_2 t} \right. \\ & \left. + K_3 e^{-\sigma t} \sin(\rho t + 2\phi_1 - \phi_2 - \phi_3)\right) v \qquad (2.16) \\ K_0 = & \frac{(\tau_1 + \tau_2)(\sigma^2 + \rho^2) + 2\sigma\tau_1\tau_2}{\tau_1\tau_2(\sigma^2 + \rho^2)} \\ K_1 = & \frac{\tau_2(\sigma^2 + \rho^2)}{\tau_1(\tau_2 - \tau_1)(\tau_1^2 - 2\sigma\tau_1 + \sigma^2 + \rho^2)} \\ K_2 = & \frac{\tau_1(\sigma^2 + \rho^2)}{\tau_2(\tau_1 - \tau_2)(\tau_2^2 - 2\sigma\tau_2 + \sigma^2 + \rho^2)} \\ K_3 = & \frac{\tau_1\tau_2}{\rho\sqrt{((\tau_1 - \sigma)^2 + \rho^2)((\tau_2 - \sigma)^2 + \rho^2)}} \end{aligned}$$

where $\phi_1 = \tan^{-1}(\rho/\sigma)$, $\phi_2 = \tan^{-1}(\rho/(\tau_1 - \sigma))$, $\phi_3 = \tan^{-1}(\rho/(\tau_2 - \sigma))$, K_0 steady-state velocity deviation of the 4th order model, K_1, K_2 response component of two real poles, K_3 response component of complex conjugate root.

(3) Relation between Servo Parameters and Characteristic Root

By using the ramp response of the normalized 4th order model, the relation between servo parameters and characteristic root is investigated. The moment of inertia ratio is given as $N_L = 3$, whose value is always adopted in industrial mechatronic servo systems.

The region of c_p and c_v satisfying conditions A, B, C is illustrated in Fig. 2.3(a),(b),(c), respectively. Fig. 2.3(d) shows the equivalent height line about the region of c_p and c_v satisfying conditions A, B, C and principal root τ_1. When the region of the response component of the complex conjugate root of condition B is very small,

$$\frac{K_3}{K_1} \leq 0.1 \tag{2.17}$$

is given. When the region of the response component of the complex conjugate root of condition C is converged quickly

$$\frac{\sigma}{\tau_1} \geq 2.0 \tag{2.18}$$

is given.

For reference, the calculated ratio of the response component K_1 of principal root when changing parameters c_p and c_v, and response component K_3 of the complex conjugate root is shown in Fig. 2.4(a). The calculated ratio of the principal root $-\tau_1$ and the real part $-\sigma$ of the complex conjugate root

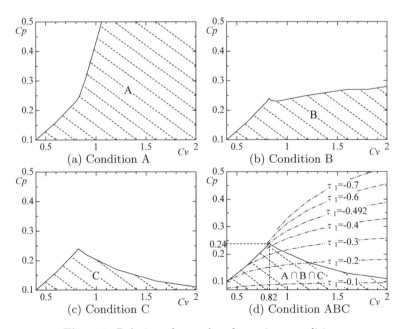

Fig. 2.3. Relation of c_p and c_v for various conditions

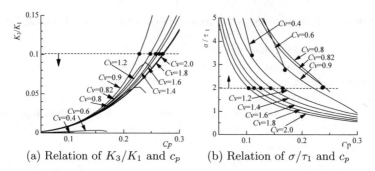

Fig. 2.4. Relation of various parameters for various c_v

is shown in Fig. 2.4(b). From Fig.(a), when c_v is fixed and c_p is increased, K_3/K_1 becomes big. That is, the response component of the complex conjugate root cannot be neglected. In Fig.(b), when c_v is fixed and c_p is increased, σ/τ_1 becomes small. That is, the declination of the response component of complex conjugate root is delayed.

(4) Determination Method of Servo Parameters Based on Control Performance

From the servo parameter determination conditions of 2.1.3(1), the servo parameters c_p and c_v are determined in order to obtain the fast response when satisfying equation (2.17) in 2.1.3(3) and equation (2.18), i.e., the principal root τ_1 is small.

According to the equivalent height line of principal root τ_1 shown in Fig. 2.3(d), when the servo parameters are $c_p = 0.24$ and $c_v = 0.82$, the minimal value is $\tau_1 = -0.492$. This is the general result which is not dependent on the natural angular frequency ω_L of the mechanism part in the normalized 4th order model (2.15).

In order to verify the obtained servo parameter results, the results of ramp response calculated by equation (2.16) are illustrated in Fig. 2.5. Fig.(a) shows the results when $N_L = 3$. Fig.(b) shows the results when $N_L = 10$. In the common velocity response of Fig.(a) and Fig.(b), the conditions of faster response in the region of no oscillation or overshoot generation are $c_p = 0.24$ and $c_v = 0.82$. In addition, by comparing the results of Fig.(a) and Fig.(b), the position and velocity are almost the same. With the general industrial field condition $3 \leq N_L \leq 10$, the conditions of faster response in velocity response without oscillation or overshoot generation are $c_p = 0.24$ and $c_v = 0.82$.

From these results, the servo parameters K_p, K_v are calculated by the natural angular frequency ω_L of the mechanism in experiment. In equation (2.12a) and (2.12b)

$$c_p = 0.24 \tag{2.19a}$$
$$c_v = 0.82 \tag{2.19b}$$

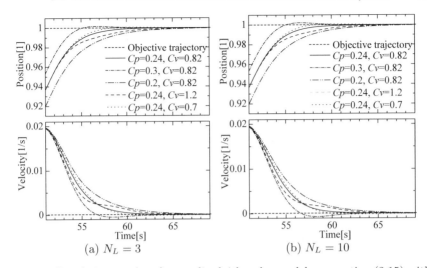

Fig. 2.5. Simulation results of normalized 4th order model as equation (2.15) with various c_p and c_v $c_p = 0.24$, $c_v = 0.82$; $c_p = 0.3$, $c_v = 0.82$; $c_p = 0.2$, $c_v = 0.82$; $c_p = 0.24$, $c_v = 1.2$; $c_p = 0.24$, $c_v = 0.7$, (a) $N_L = 3$, (b) $N_L = 10$.

are given. The regulation of a mechatronic servo system, for fast response without oscillation or overshoot, can be carried out.

2.1.4 Experiment Verification of the Mathematical Model

(1) Simulation and Experiment

The appropriation of the determination method for the servo parameter of industrial mechatronic servo system, derived in the former part, is verified by the experiment of DEC-1 (refer to experiment device E.1). The sampling time interval of the experiment is given as 1[ms] (refer to 3.1). The value of position loop gain K_p can be changed in the computer program. The value of velocity loop gain K_v needs the equivalent value when K_v^s in Fig. 1.1 is adjusted by altering the variable resistance. The concrete method is that, when the position loop is at the outside and the step signal of velocity is given, the time constant corresponding to this response wave is worked out and K_v is calculated by its inverse value. When changing the value of variable resistance, the variable resistance, as the regulation value, which is consistent with the determined K_v value by the above experiment with the method of 2.1.3(4), is adopted. With this method, the ratio gain K_v^s of the PI controller of the actual velocity controller, corresponding to the optimal gain K_v of P controller of velocity control in the 4th order model, can also be worked out.

The motion velocity of the mechatronic servo system serves as the operation velocity in the general industrial field. With about 1/10 of motor rated speed $u(t) = 10t$[rad/s] as well as two conditions (a) K_p=22.6[1/s],

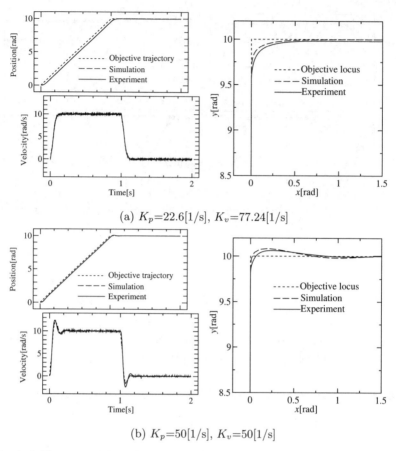

Fig. 2.6. Experimental results by using DEC-1 experiment device and comparison with simulation results by using 4th order model

$K_v=77.24[1/\text{s}]$, (b) $K_p=50[1/\text{s}]$, the experiment was carried out. Condition (a) is the appropriate servo parameter calculated by putting $c_p = 0.24$, $c_v = 0.82$ and $\omega_L = 94.2[\text{rad/s}]$ into equation (2.12a) and (2.12b). Condition (b) is the deviation of the servo parameter from the proper value. These experimental results and simulation results are illustrated in Fig. 2.6. However, for grasping visually the influence given to contour control performance, Fig. 2.6 shows the expansion graph of the angular part of the contour control results when same experimental results were used twice for the positions of the x axis and the y axis.

For proper servo parameters and servo parameters complete with errors, the simulation results based on the 4th order model of a mechatronic servo system are almost identical to the actual experimental results. Therefore, it verified that the 4th order model is the correct expression of the dynamic characteristic of an industrial mechatronic servo system. The validation of adaptation of the 4th order model in the design of a servo controller is also shown.

Moreover, in the simulation and experimental result of condition (a), the desired response characteristics without oscillation or overshoot at all in both position response and velocity response is illustrated. However, in condition (b), the position response is near to the objective trajectory comparing with that of (a). But oscillation is generated both in the position response and velocity response. Additionally, in contour control, the overshoot has occurred and the control performance has deteriorated. Since this overshoot must be avoided in the contour control in the industrial field, this condition cannot be adopted in the contour control. Based on the above explanation, the effectiveness of the proposed determination method of servo parameter was verified by experimental results.

In an industrial mechatronic servo system, for regulating each axis characteristic with consistence, this method is adapted for all axes of the mechatronic servo system and the high-precision contour control of industrial mechatronic servo systems can be realized.

2.2 Reduced Order Model of One Axis in a Mechatronic Servo System

The expression of a mechatronic servo system by a reduced order model corresponding to the movement velocity condition is desired from the simple controller design.

According to the 4th order model, the model approximation error is defined and the linear 1st order equation (2.23) and the linear 2nd order equation (2.29) are constructed. The relation between the model parameters of the 4th order model and the model parameters of the reduced order model is given in equations (2.24), (2.30) and (2.37).

The 1st order model for expressing the low speed operation of the mechatronic servo system (velocity below 1/20 rated speed) and the 2nd order model for expressing the middle speed operation (velocity below 1/5 rated speed) trace the experience of one of the authors. The significance of these reduced order models has been proved. The effective usage of the model for servo controller design is also verified by example.

2.2.1 Necessary Conditions of the Reduced Order Model

As introduced in section 2.1, one axis of mechatronic servo system is constructed by many blocks (parts). These blocks (parts) have respectively at least one or two order transfer functions. From block diagrams expressing correctly these blocks, it is very difficult to grasp quickly and entirely the features of the servo system. In an industrial field, these mechatronic servo systems are previously regarded as a simple 1st order system (refer to 1.2.1(1)). However, since these are the approximated judgment from the movement of the mechatronic servo system, it is hard to say that this possesses the distinctly theoretical ground.

In this section, considering the selection method of the servo motor firstly, the necessity of the reduced order model of the mechatronic servo system is arranged as below.

1. In the mechanism part determined from the operation purpose (the features of mechanism part are expressed by natural angular frequency and damping rate), the servo motor is set up according to the motor selection method [8]. When controlling this servo motor by the servo controller, the actual mechanism is established according to the whole features of the servo system and the entire servo system is known before regulation.
2. For understanding the entire features, the exchange of the mechanism part is needed and also the revision of motor selection should be judged.
3. From this feature, it should judge how long to follow the current assumed operation pattern (Generally, trapezoidal wave of velocity is always adopted in the positioning control).
4. In the contour control, the trace of actual trajectory in term of command should be judged and the proper action should be briefly known.

Next, the important factors in the reduced order model are listed below.

1. The features of the main structure blocks of the mechatronic servo system (such as natural angular frequency of the mechanism part, properties of damping rate and motor, etc) should be reflected.
2. The general regulation condition of the servo system (overshoot is not absolutely generated not only in the position loop but also in the velocity loop) should be reflected.
3. The action conditions of the servo system (e.g., the instruction is the ramp input of each independent axis, the trajectory speed in the contour control is below 1/5 of maximum velocity, etc) should be reflected.
4. The reduced order is adopted for modeling and one model can be used for one action status.

The reduced order model of mechatronic servo systems satisfying the above conditions is the 1st order model in low speed contour control, i.e., the characteristic parameter is only K_{p1}; the 2nd order model in middle contour control, i.e., the characteristic parameters are K_{p2}, K_{v2}. The detailed explanation is as below.

2.2.2 Structure Standard of Model

With the 4th order model (2.13) as standard, for the contour control of industrial mechatronic servo systems, low speed 1st order model expressing properly

2.2 Reduced Order Model of One Axis in a Mechatronic Servo System

the 1/20 of rated speed and middle speed 2nd order model expressing properly the system with the speed from 1/20 of rated speed to 1/5 of rated speed are constructed. Concerning the above velocities, from the nonlinear feature in the control system, especially the effect of torque saturation, modeling is very complicated. Moreover, from this nonlinear feature, if the contour control cannot be carried out for position determination, modeling is not needed for contour control.

The structure standard of the reduced order model is determined by the following conditions based on the 4th order model expressing by equation (2.13).

1. The steady-state velocity deviation between the 4th order model and the reduced order model are consistent.
2. The oscillation does not occur in the ramp response of the reduced order model.
3. The squared integral of the ramp response error between the 4th order model and the reduced order model is minimized.

Regarding the ramp response as standard is to agree with the actual application that in the contour control in industrial applications there are many kinds of motion with a constant trajectory velocity.

2.2.3 Derivation of Low Speed 1st Order Model

With the movement velocity smaller than 1/20 of rated speed, the low speed 1st order model expressing properly the industrial mechatronic servo system can be derived. This low speed 1st order model is expressed as a 1st order system. In the mechanism part, the inertial moment of the load is transformed into the motor axis. Considering both the whole inertial moment of the mechatronic servo system and the electric characteristic of the servo motor, the whole mechatronic servo system is as

$$\frac{dy(t)}{dt} = -c_{p1}\{y(t) - u(t)\} \quad (2.20)$$

and its model expressed by transfer function is as

$$G_{c1}(s) = \frac{c_{p1}}{s + c_{p1}} \quad (2.21)$$

where the relation of parameter c_{p1} and the position loop gain K_{p1} of the low speed 1st order model (refer to Fig. 2.7) is as

$$K_{p1} = c_{p1}\omega_L. \quad (2.22)$$

The low speed 1st order model as equation (2.21) is the model independent of the load natural angular frequency ω_L, as similar with the normalized 4th order model as equation (2.13). That is, if given the natural angular frequency

ω_L, the low speed 1st order model can be derived corresponding to the ω_L of equation (2.21), (2.22). The transfer function $G_1(s)$ of the low speed 1st order model without normalization by using position loop gain K_{p1} is as

$$G_1(s) = \frac{K_{p1}}{s + K_{p1}}. \qquad (2.23)$$

When equation (2.22) is put into equation (2.21), the form is changed by revising $s\omega_L$ with s. That is, the scale of time axis is transformed from t/ω_L to t.

The parameter c_{p1} in the low speed 1st order model (2.21) can be derived with the condition 1 of 2.2.2 and for agreement with the steady-state velocity deviation as

$$c_{p1} = \frac{b_0}{b_1} \approx c_p, \qquad (2.24)$$

Here, the final approximation equation in (2.24) is the results approximated with $\zeta_L \approx 0$ for very small damping rate from 0 to 0.02 of the mechanism part in the industrial mechatronic servo system. When given $c_p = 0.24$ in the mechatronic servo system regulated properly, $c_{p1} = 0.24$ is better to be given for approximation of equation (2.24).

2.2.4 Derivation of the Middle Speed 2nd Order Model

Next, the middle speed 2nd order model expressing properly the industrial mechatronic servo system from 1/20 to 1/5 of rated speed can be derived. This middle speed 2nd order model is the 2nd order system. The whole mechatronic servo system is as

$$\frac{d^2 y(t)}{dt^2} = -c_{v2}\frac{dy(t)}{dt} - c_{p2}c_{v2}y(t) + c_{p2}c_{v2}u(t) \qquad (2.25)$$

and the model expressing by transfer function is as

$$G_{c2}(s) = \frac{c_{v2}c_{p2}}{s^2 + c_{v2}s + c_{v2}c_{p2}}. \qquad (2.26)$$

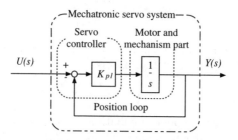

Fig. 2.7. Low speed 1st order model of industrial mechatronic servo system

2.2 Reduced Order Model of One Axis in a Mechatronic Servo System

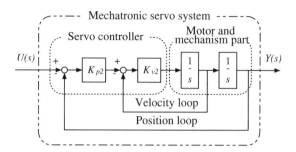

Fig. 2.8. Middle speed 2nd order model of industrial mechatronic servo system

Here, the relationship between the parameter c_{p2}, c_{v2}, position loop gain K_{p2} and velocity loop gain K_{v2} of the middle speed 2nd order model (refer to Fig. 2.8) are as

$$K_{p2} = c_{p2}\omega_L \tag{2.27}$$
$$K_{v2} = c_{v2}\omega_L. \tag{2.28}$$

That is, if given the natural angular frequency ω_L, the middle speed 2nd order model corresponding to the ω_L in equation (2.26), (2.27) and (2.28) can be derived. As same as the low speed 1st order model, the transfer function $G_2(s)$ of the middle speed 2nd order model without normalization by using position loop gain K_{p2} and velocity loop gain K_{v2} is as

$$G_2(s) = \frac{K_{v2}K_{p2}}{s^2 + K_{v2}s + K_{v2}K_{p2}}. \tag{2.29}$$

From the condition 1 of item 2.2.2 and for agreement with the steady-state velocity error, the parameter c_{p2} and c_{v2} in middle speed 2nd order model (2.26) is as

$$c_{p2} = \frac{b_0}{b_1} \approx c_p. \tag{2.30}$$

Next, analyzing conditions 2 and 3 in item 2.2.2, the squared integral of the model output error between the normalized 4th order model and the middle speed 2nd order model is derived.

If the 2nd order model (2.26) is expressed as

$$G_{c2} = \frac{\omega_2^2}{s^2 + 2\zeta_2\omega_2 s + \omega_2^2} \tag{2.31}$$

$$c_{p2} = \frac{\omega_2}{2\zeta_2}$$

$$c_{v2} = 2\zeta_2\omega_2$$

from the condition 2 of item 2.2.2, the condition of no oscillation generation in the response of the 2nd order model is firstly considered as $\zeta_2 > 1$ for satisfying

$\zeta_2 \geq 1$. When $\zeta_2 > 1$, i.e., there are two real poles p_1, p_2, the response of the 2nd order model is as below with the ramp input $u = vt$ from equation (2.26).

$$y_2(t) = \left(t - \frac{2\zeta_2}{\omega_2} + \frac{(\zeta_2 - \sqrt{\zeta_2^2 - 1})e^{p_1 t}}{2\omega_2(1 - \zeta_2^2 - \zeta_2\sqrt{\zeta_2^2 - 1})} \right.$$
$$\left. + \frac{(\zeta_2 + \sqrt{\zeta_2^2 - 1})e^{p_2 t}}{2\omega_2(1 - \zeta_2^2 + \zeta_2\sqrt{\zeta_2^2 - 1})} \right) v \qquad (2.32)$$

where, $p_1 = -(\zeta_2 + \sqrt{\zeta_2^2 - 1})\omega_2$ and $p_2 = -(\zeta_2 - \sqrt{\zeta_2^2 - 1})\omega_2$. When we put $K_0(= b_1/b_0) = 2\zeta_2/\omega_2(= 1/c_{p2})$, which is the equivalent condition of the velocity steady-state deviation between the normalized 4th order model and the 2nd order model, into the equation (2.32), the squared integral of the model output error between the normalized 4th order model, which is from the ramp response (2.16) of relationship $\omega_2 = 2\zeta_2 c_{p2}$ and normalized 4th model, and the 2nd order model is given as

$$J_2 = \left(\frac{(\tau_1 + \tau_2)(K_1^2 \tau_2 + K_2^2 \tau_1) + 4K_1 K_2 \tau_1 \tau_2}{2\tau_1\tau_2(\tau_1 + \tau_2)} \right.$$
$$+ \frac{16\zeta_2^4 - 4\zeta_2^2 + 1}{32 c_{p2}^3 \zeta_2^4} - \frac{2K_1((\tau_1 - c_{p2})\zeta_2 + 4c_{p2}\zeta_2^3)}{c_{p2}\tau_1^2 \zeta_2 + 4c_{p2}^2(\tau_1 + c_{p2})\zeta_2^3}$$
$$\left. - \frac{2K_2((\tau_2 - c_{p2})\zeta_2 + 4c_{p2}\zeta_2^3)}{c_{p2}\tau_2^2 \zeta_2 + 4c_{p2}^2(\tau_2 + c_{p2})\zeta_2^3} \right) v^2. \qquad (2.33)$$

The squared integral of the output error between the normalized 4th order model and the 2nd order model is calculated with the differential about ζ_2 by equation (2.33) as

$$\frac{dJ_2}{d\zeta_2} = \left(\frac{2\zeta_2^2 - 1}{8c_{p2}^3 \zeta_2^5} + \frac{16K_1 c_{p2}^4 \zeta_2^3}{(c_{p2}\tau_1^2 \zeta_2 + 4c_{p2}^2(\tau_1 + c_{p2})\zeta_3^3)^2} \right.$$
$$\left. + \frac{16K_2 c_{p2}^4 \zeta_2^3}{(c_{p2}\tau_2^2 \zeta_2 + 4c_{p2}^2(\tau_2 + c_{p2})\zeta_2^3)^2} \right) v^2. \qquad (2.34)$$

This value is often positive if $\zeta_2 > 1$. That is, since $J_2(\zeta_2)$ is the mono-increase function in the scale of $\zeta_2 > 1$, $J_{2\min} = \lim_{\zeta_2 \to 1} J_2(\zeta_2)$. If $\zeta_2 = 1$, the squared integral of the output error between the normalized 4th order model and the 2nd order model is given with a minimum value. If $\zeta_2 = 1$ then $c_{p2} = \omega_2/2$ and $c_{v2} = 2\omega_2$. Its result is $c_{v2} = 4c_{p2}$. In addition, its ramp response of the 2nd order model is

$$y_2(t) = \left(t - \frac{1}{c_{p2}} + \left(t + \frac{1}{c_{p2}} \right) e^{-2c_{p2} t} \right) v. \qquad (2.35)$$

Besides, the minimal value of the squared integral of the output error between the normalized 4th order model and the 2nd order model is calculated as

2.2 Reduced Order Model of One Axis in a Mechatronic Servo System

Table 2.1. Evaluation of reduced order model (rated speed $V_M = 104[\text{rad/s}]$, $\omega_L = 94.2[\text{rad/s}]$, servo parameter of low speed 1st order model $K_{p1} = 23.6[1/\text{s}]$, servo parameter of middle speed 2nd order model $K_{p2} = 23.6[1/\text{s}]$, $K_{v2} = 84.8[1/\text{s}]$)

Velocity [rad/s]	Low velocity eq(2.23)[rad²]		Middle velocity eq(2.29)[rad²]	
$5.02(= V_M/20)$	7.07×10^{-5}	◯	5.18×10^{-5}	◯
$20.1(= V_M/5)$	1.13×10^{-3}	×	8.30×10^{-5}	◯
$34.0(= V_M/3)$	7.07×10^{-3}	×	5.18×10^{-4}	×

$$J_{2\min} = \left(\frac{(\tau_1 + \tau_2)(K_1^2 \tau_2 + K_2^2 \tau_1) + 4K_1 K_2 \tau_1 \tau_2}{2\tau_1 \tau_2 (\tau_1 + \tau_2)} \right.$$
$$\left. + \frac{13}{32 c_{p2}^3} - \frac{2K_1(\tau_1 + 3c_{p2})}{c_{p2}(\tau_1 + 2c_{p2})^2} - \frac{2K_2(\tau_2 + 3c_{p2})}{c_{p2}(\tau_2 + 2c_{p2})^2} \right) v^2. \tag{2.36}$$

From the above discussion, c_{v2} satisfying conditions can be derived for the minimum by

$$c_{v2} = 4c_{p2} \approx 4c_p. \tag{2.37}$$

The approximation equation (2.30) is as same as (2.24). The approximation equation (2.37) uses the approximation equation of (2.30). In the mechatronic servo system regulated properly, $c_p = 0.24$ is given. From equation (2.30) and (2.37), $c_{p2} = 0.24$ and $c_{v2} = 0.96$ are given.

2.2.5 Evaluation of the Low Speed 1st Order Model and the Middle Speed 2nd Order Model

Through the respective movement velocities of the low speed 1st order model and the middle speed 2nd order model derived in 2.2.3 and 2.2.4, the appropriate modeling mechatronic servo system is illustrated. In the contour control of an industrial mechatronic servo system, ramp input is always adopted. As the performance standard of the reduced order model, the error squared integral of the ramp response error between the 4th order model and the reduced order model is adopted.

In the contour control, the ramp input of mechatronic servo system is 1/20 of the maximum in the scale of motor rated speed from 1/100 to 1/20, or 1/5 of maximum in that of rated speed from 1/20 to 1/5, or 1/3 of maximum in that of rated speed from 1/5 to 1/3. The calculation results of the squared integral of the model output error between the reduced order model and the normalized 4th order model are illustrated in table 2.1. If given the allowance error $1 \times 10^{-4}[\text{rad}^2]$, the symbol ◯ in the table denotes satisfying the allowance error and × denotes not satisfying the allowance error.

From the table 2.1, in the low speed operation from 1/100 to 1/20 of the rated speed of the motor, the evaluation error between the low speed 1st order model and the middle speed 2nd order model is smaller than the required allowable error. When constructing the model, in order to obtain the simple model with satisfying the required precision, the model should be the low speed 1st order model for the low speed operation. Additionally, in the middle speed operation from 1/20 to 1/5 of rated speed, the evaluation error of the low speed 1st order model is bigger than the required allowance error and smaller than that in the middle speed 2nd order model. In the high-speed motion over 1/5 of rated speed of the motor, the evaluation error between the low speed 1st order model and middle speed 2nd order model is bigger than the required allowance error. From these results, the adaptable scale and boundary of the reduced order model can be judged.

The correct modeling of actual industrial mechatronic servo system by derived reduced order model was verified by experiment. The adopted experimental device for verification is a DEC-1 similar to item 2.1.3 (refer to experimental device E.1). The low speed of motion velocity is 5[rad/s] about 1/20 of rated speed, and middle speed is 20[rad/s] about 1/5 of rated speed. Fig. 2.9 illustrates the modeling error between the output and the reduced order model in the experiment. From the results in Fig. 2.9, in the low speed operation, the modeling error of both the low speed 1st order model and the middle speed 2nd order model is smaller than 0.05[rad], which is almost consistent with the experimental results. In the middle speed operation, the error between the low speed 1st order model and experimental results is bigger than the maximal 0.14[rad]. In the middle speed 2nd order model, the modeling error is smaller than 0.05[rad]. Therefore, the modeling is appropriate. From these experimental results, the appropriateness of the reduced order model expressing the dynamic of industrial mechatronic servo system was verified.

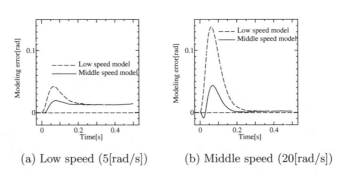

(a) Low speed (5[rad/s])   (b) Middle speed (20[rad/s])

Fig. 2.9. Evaluation of low speed 1st order model and middle speed 2nd order model

2.3 Linear Model of the Working Coordinates of an Articulated Robot Arm   37

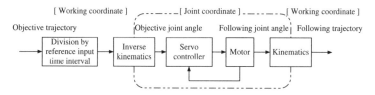

Fig. 2.10. Block diagram of industrial articulated robot arm

2.3 Linear Model of the Working Coordinates of an Articulated Robot Arm

In an industrial articulated robot arm, instructions are given in working coordinate. The motor is driven in the joint coordinate space transformed by nonlinear coordinates by calculation in the controller. Hence, the mechanism part is moved in the working coordinate space. Therefore, according to the special region in working coordinates, there is the problem of precision deterioration of the contour control of robot arm.

The approximation model (2.46) in the working coordinate of an articulated robot arm and its approximation error (2.54) are derived.

By using this model, the working linearizable approximation possible region for keeping the movement precision of an articulated robot arm within the allowance is clarified. The region, in which the high-precision contour control of the robot arm is capable to realize, is confirmed. Besides, from the discussion in this section, by holding this view of approximation error, the one axis characteristic in the joint coordinate given in 2.1 and 2.2 can express the characteristics of the mechatronic servo system in working coordinates. The simplification of the analysis and design of mechatronic servo systems is very important.

2.3.1 A Working Linearized Model of an Articulated Robot Arm

(1) An Industrial Articulated Robot Arm Control System

The block diagram of contour control of an industrial articulated robot arm is illustrated in Fig. 2.10. At first, the objective trajectory in working coordinates is divided into each reference input time interval (refer to section 3.2 and 3.3). The joint angle of each axis is calculated at each division point. The rotation angle of the servo motor is controlled by various axis joint angles with constant velocity movements based on the objective joint angle divided in joint coordinate. The servo motor of each axis is rotated only with its defined movement. Thus, the arm tip is moved along the objective trajectory of the working coordinate with the coordinate transform in the arm mechanism.

If the objective trajectory is given in working coordinates and the robot arm control of each axis is independent of the joint coordinate with nonlinear transform, the following trajectory is evaluated in working coordinates with nonlinear transform. When controlling a robot arm with this control pattern, the control system of an industrial robot arm, with each linear independent coordinate axis, is generally approximated in working coordinates. For preparing the discussion (in 2.3.2) of appropriate linear approximation in this working coordinate, the working linearized approximation trajectory, based on the actual trajectory and working linearized model of working coordinate of this robot arm control system, is derived.

(2) Actual Trajectory of a Two-Axis Robot Arm

For analyzing the characteristics of multiple axes, the nature of two axes is discussed and the analysis is expanded into multiple axes in 2.3.2(4). In Fig. 2.11, two rigid links are expressed with ⊙. The conceptual graph of a two-axis robot arm with movement of the tip on this plate is shown. The (θ_1, θ_2) in figure is the joint angle in joint coordinates. (p_x, p_y) is the tip position in working coordinates, l_1, l_2 are the lengths of axis 1 and axis 2, respectively. This two-axis robot arm is the basic structure of a multi-axis robot arm. In the SCARA robot arm, the plate position determination is carried out for these two axes.

At first, for determining the relationship between the working coordinate and joint coordinate, the transformation from joint coordinate (θ_1, θ_2) to working coordinate (p_x, p_y) (kinematics) and the transformation from working coordinate (p_x, p_y) to joint coordinate (θ_1, θ_2) are explained. From Fig. 2.11, the kinematics is as

$$p_x = l_1 \cos\theta_1 + l_2 \cos(\theta_1 + \theta_2) \tag{2.38a}$$
$$p_y = l_1 \sin\theta_1 + l_2 \sin(\theta_1 + \theta_2). \tag{2.38b}$$

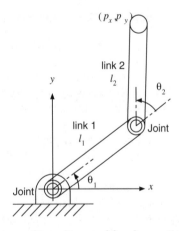

Fig. 2.11. Structure of two-degree-of-freedom articulated robot arm

2.3 Linear Model of the Working Coordinates of an Articulated Robot Arm

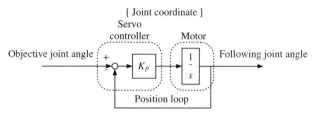

Fig. 2.12. Block diagram of 1st order model in joint coordinate of industrial mechatronic servo system

From the solution of (θ_1, θ_2) in equation (2.38a) and (2.38b), the inverse kinematics is given as

$$\theta_1 = \sin^{-1}\left(\frac{p_y}{\sqrt{p_x^2 + p_y^2}}\right) - \sin^{-1}\left(\frac{l_2 \sin\theta_2}{\sqrt{p_x^2 + p_y^2}}\right) \quad (2.39a)$$

$$\theta_2 = \pm\cos^{-1}\left(\frac{p_x^2 + p_y^2 - l_1^2 - l_2^2}{2l_1 l_2}\right) \quad (2.39b)$$

where the symbol of equation (2.39b) denotes that one assigned point in working coordinate has two possibilities in the joint coordinate.

Next, the dynamics of the robot arm is given in the joint coordinate. In an industrial robot arm, if the gear ratio is large, then the load inertia is small. Moreover, if using a parallel link, the effect of no-angle part of inertia matrix is small. The servo motor in the actuator performs the control on the robot arm in each independent axis. For an actual industrial robot arm, when the motion velocity of the robot arm is below 1/20 of rated speed, each axis can be expressed with a 1st order system as (refer to 2.2.3).

$$\frac{d\theta_1(t)}{dt} = -K_p\theta_1(t) + K_p u_1(t) \quad (2.40a)$$

$$\frac{d\theta_2(t)}{dt} = -K_p\theta_2(t) + K_p u_2(t). \quad (2.40b)$$

The model expressed by equation (2.40) is called a joint linearized model. Here, $u_1(t)$ and $u_2(t)$ denotes the angle input of axis 1 and axis 2, respectively. K_p denotes K_{p1} of equation (2.23) in the low speed 1st order model of 2.2.3. Fig. 2.12 illustrates the block diagram of the 1st order system. In this section, each axis dynamic is expressed by equation (2.40) in joint coordinates. For clarifying the expression of actual robot dynamics by the joint linearized model. The following discussion is carried out with this assumption.

The robot arm is analyzed about how to trace the objective trajectory divided into small intervals. Concerning the various trajectories divided from the objective trajectory, the beginning point and end point in working coordinates within one divided small interval are expressed by (p_x^0, p_y^0), $(p_x^{\Delta T}, p_y^{\Delta T})$,

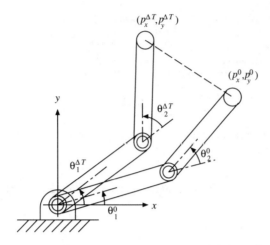

Fig. 2.13. One interval of objective trajectory divided by reference input time interval

respectively, and the beginning point and end point in joint coordinates are expressed by (θ_1^0, θ_2^0), $(\theta_1^{\Delta T}, \theta_2^{\Delta T})$, respectively. The relationship between joint coordinates and working coordinates in this small interval is given in Fig. 2.13. The relation between (p_x^0, p_y^0) and (θ_1^0, θ_2^0) as well as between $(\theta_1^{\Delta T}, \theta_2^{\Delta T})$ and $(p_x^{\Delta T}, p_y^{\Delta T})$ are expressed as below based on the expression of the relationship between working coordinates and joint coordinates from equation (2.38a) and (2.38b).

$$p_x^0 = l_1 \cos \theta_1^0 + l_2 \cos(\theta_1^0 + \theta_2^0) \tag{2.41a}$$
$$p_y^0 = l_1 \sin \theta_1^0 + l_2 \sin(\theta_1^0 + \theta_2^0) \tag{2.41b}$$
$$p_x^{\Delta T} = l_1 \cos \theta_1^{\Delta T} + l_2 \cos(\theta_1^{\Delta T} + \theta_2^{\Delta T}) \tag{2.41c}$$
$$p_y^{\Delta T} = l_1 \sin \theta_1^{\Delta T} + l_2 \sin(\theta_1^{\Delta T} + \theta_2^{\Delta T}). \tag{2.41d}$$

Concerning the industrial robot arm, from the given constant angle velocity input (v_1, v_2) of each axis in divided small intervals, the angle input $(u_1(t), u_2(t))$ for each axis dynamic of the robot arm (2.40) is given as $(u_1(t), u_2(t))$

$$u_1(t) = \theta_1^0 + v_1 t, \quad v_1 = \frac{\theta_1^{\Delta T} - \theta_1^0}{\Delta T} \tag{2.42a}$$

$$u_2(t) = \theta_2^0 + v_2 t, \quad v_2 = \frac{\theta_2^{\Delta T} - \theta_2^0}{\Delta T} \tag{2.42b}$$

where ΔT denotes the reference input time interval (refer to 3.2, 3.3). The time of the beginning division point is zero.

If the angle input is expressed by equation (2.42), the robot arm position in working coordinates can be derived. When the objective trajectory is the

2.3 Linear Model of the Working Coordinates of an Articulated Robot Arm

same as the position of the actual trajectory as $(\theta_1(0), \theta_2(0)) = (\theta_1^0, \theta_2^0)$ in the initial time of robot arm, the position in joint coordinates of the robot arm is as below from the solution of differential equation after putting angle input of equation (2.42) into (2.40) (refer to appendix A.2).

$$\theta_1(t) = \theta_1^0 + v_1 \lambda(t) \qquad (2.43a)$$
$$\theta_2(t) = \theta_2^0 + v_2 \lambda(t) \qquad (2.43b)$$

$$\lambda(t) = t + \frac{e^{-K_p t} - 1}{K_p}. \qquad (2.44)$$

At this time, the position of the robot arm in working coordinate can be calculated when putting the nonlinear transform equation (2.43) into (2.38a), (2.38b)

$$p_x(t) = l_1 \cos\{\theta_1^0 + v_1 \lambda(t)\} + l_2 \cos\{\theta_1^0 + \theta_2^0 + (v_1 + v_2)\lambda(t)\} \qquad (2.45a)$$
$$p_y(t) = l_1 \sin\{\theta_1^0 + v_1 \lambda(t)\} + l_2 \sin\{\theta_1^0 + \theta_2^0 + (v_1 + v_2)\lambda(t)\}. \qquad (2.45b)$$

This equation (2.45) expresses the actual trajectory of the robot arm tip in working coordinates. Concerning this actual trajectory, as the problem of this section, the working linearized approximation trajectory in the working linearized model is derived after linearized approximation of each coordinate axis independently of the working coordinates.

(3) Working Linearized Approximation Trajectory of a Two-Axis Robot Arm

In working coordinates, the control system of the robot arm is as below when x axis y axis are linearly approximated independently, respectively

$$\frac{d\hat{p}_x(t)}{dt} = -K_p \hat{p}_x(t) + K_p u_x(t) \qquad (2.46a)$$
$$\frac{d\hat{p}_y(t)}{dt} = -K_p \hat{p}_y(t) + K_p u_y(t) \qquad (2.46b)$$

where $(\hat{p}_x(t), \hat{p}_y(t))$ denotes the robot arm position in the working coordinate linearly approximation. $(u_x(t), u_y(t))$ denotes the position input in working coordinates. This equation (2.46) is the working linearized model as the discussion object of this section. When the objective trajectory is divided as shown in Fig. 2.13 with the linearized approximation equation (2.46), the robot arm response at small intervals is derived. Here, the objective trajectory is the same as the position of the working linearized approximation trajectory as $(\hat{p}_x(0), \hat{p}_y(0)) = (p_x^0, p_y^0)$ at the initial time of the robot arm. Strictly speaking, The input in working coordinate corresponding to the input (2.42) in the joint coordinate needs to be derived according to the coordinate transform (2.38a), (2.38b). If the input in the working coordinate is not a constant velocity, the input in working coordinates is approximated with a constant velocity by

$$u_x(t) = p_x^0 + v_x t, \quad v_x = \frac{p_x^{\Delta T} - p_x^0}{\Delta T} \tag{2.47a}$$

$$u_y(t) = p_y^0 + v_y t, \quad v_y = \frac{p_y^{\Delta T} - p_y^0}{\Delta T}. \tag{2.47b}$$

Its approximation error can almost be neglected. If the input of the equation (2.47) is put into the working linearized model of equation (2.46), the working linearized approximation trajectory of the robot arm from the solution of differential equation is as

$$\hat{p}_x(t) = p_x^0 + v_x \lambda(t) \tag{2.48a}$$
$$\hat{p}_y(t) = p_y^0 + v_y \lambda(t). \tag{2.48b}$$

That is, the working linearized approximation trajectory corresponding to the actual trajectory (2.45) of the robot arm in working coordinates is given by equation (2.48).

2.3.2 Derivation of Adaptable Region of the Working Linearized Model

(1) Approximation Error of the Working Linearized Model

From the comparison between the actual trajectory (2.45) of the robot arm control system and the working linearized approximation trajectory (2.48), the approximation precision of the working linearized model for the object discussed in this section is evaluated. The approximation error in the working coordinate is the error between equation (2.45) and (2.48) as

$$e_x(t) = \hat{p}_x(t) - p_x(t) \tag{2.49a}$$
$$e_y(t) = \hat{p}_y(t) - p_y(t). \tag{2.49b}$$

$(e_x(t), e_y(t))$ of equation (2.49) is called the working linearized approximation error. In order to evaluate separately the item about the time and the item about the space in equation (2.49), the actual position of the robot arm in working coordinates expressed by equation (2.45) is calculated as below with 1st order approximation by Taylor expansion when the movement of (θ_1^0, θ_2^0) is very small.

$$\begin{aligned}\tilde{p}_x(t) = &\; l_1\{\cos(\theta_1^0) - \sin(\theta_1^0) v_1 \lambda(t)\} \\ &+ l_2\{\cos(\theta_1^0 + \theta_2^0) - \sin(\theta_1^0 + \theta_2^0)(v_1 + v_2)\lambda(t)\} \end{aligned} \tag{2.50a}$$

$$\begin{aligned}\tilde{p}_y(t) = &\; l_1\{\sin(\theta_1^0) + \cos(\theta_1^0) v_1 \lambda(t)\} \\ &+ l_2\{\sin(\theta_1^0 + \theta_2^0) + \cos(\theta_1^0 + \theta_2^0)(v_1 + v_2)\lambda(t)\}. \end{aligned} \tag{2.50b}$$

2.3 Linear Model of the Working Coordinates of an Articulated Robot Arm

Between the actual trajectory and the 1st order approximation trajectory by Taylor expansion of equation (2.50) is as [9]

$$p_x(t) = \tilde{p}_x(t) + l_1 o\{v_1 \lambda(t)\} + l_2 o\{(v_1 + v_2)\lambda(t)\}$$
$$= \tilde{p}_x(t) + o\{\lambda(t)\} \quad (2.51a)$$
$$p_y(t) = \tilde{p}_y(t) + l_1 o\{v_1 \lambda(t)\} + l_2 o\{(v_1 + v_2)\lambda(t)\}$$
$$= \tilde{p}_y(t) + o\{\lambda(t)\}. \quad (2.51b)$$

The $o\{\lambda(t)\}$ in equation (2.51) denotes the high level infinitesimal of $\lambda(t)$. By using triangle inequality, the size of error between the actual trajectory and the working linearized approximation trajectory can be restrained by equation (2.48) and (2.50)

$$|\hat{p}_x(t) - p_x(t)| \leq |\hat{p}_x(t) - \tilde{p}_x(t)| + |\tilde{p}_x(t) - p_x(t)|$$
$$= |\varepsilon_x \lambda(t)| + |o\{\lambda(t)\}| \quad (2.52a)$$
$$|\hat{p}_y(t) - p_y(t)| \leq |\hat{p}_y(t) - \tilde{p}_y(t)| + |\tilde{p}_y(t) - p_y(t)|$$
$$= |\varepsilon_y \lambda(t)| + |o\{\lambda(t)\}| \quad (2.52b)$$

where $(\varepsilon_x, \varepsilon_y)$ is

$$\varepsilon_x = v_x + p_y^0 v_1 + l_2 \sin(\theta_1^0 + \theta_2^0) v_2 \quad (2.53a)$$
$$\varepsilon_y = v_y - p_x^0 v_1 - l_2 \cos(\theta_1^0 + \theta_2^0) v_2. \quad (2.53b)$$

If the position of the robot arm is depended on velocity, there has the error item not depended on the time. When $\lambda(t)$ is very small, the item $o\{\lambda(t)\}$ in equation (2.51) can be neglected. Therefore, the working linearized approximation error can be approximated as

$$e_x(t) \approx \varepsilon_x \lambda(t) \quad (2.54a)$$
$$e_y(t) \approx \varepsilon_y \lambda(t). \quad (2.54b)$$

That is, if the $\lambda(t)$ can be very small and the division interval of the objective trajectory is very small, the working linearized approximation error can be expressed by equation (2.54). The equation (2.54) is given by item $(\varepsilon_x, \varepsilon_y)$ depended on the robot arm position in equation (2.53) and the integral with item $\lambda(t)$ dependent on time. The $(\varepsilon_x, \varepsilon_y)$ in equation (2.53) is the function of the robot arm position (p_x^0, p_y^0), (θ_1^0, θ_2^0) and motion velocity (v_x, v_y), (v_1, v_2). Here, the robot arm position (θ_1^0, θ_2^0) expressed in joint coordinates can be expressed in working coordinates by kinematic equation (2.38a), (2.38b). Moreover, the motion velocity in joint coordinates, expressed by $(v_1, v_2) = ((\theta_1^{\Delta T} - \theta_1^0)/\Delta T, (\theta_2^{\Delta T} - \theta_2^0)/\Delta T)$ in equation (2.42), can be also expressed in working coordinates as (p_x^0, p_y^0), $(p_x^{\Delta T}, p_y^{\Delta T})$ from kinematics (2.38a), (2.38b). Equation (2.54) can express the robot arm position (p_x^0, p_y^0), $(p_x^{\Delta T}, p_y^{\Delta T})$ in working coordinates. This equation (2.54) expresses the working linearized approximation error, as the purpose. From the evaluating the size of this error, the appropriation of the working linearized model of the control system of the robot arm as well as the working linearizable approximation possible region can be derived.

(2) Quantity Evaluation of the Working Linearized Model

The small region of working linearized approximation error of the working linearized model as (2.46) in working coordinates of robot arm, i.e., working linearizable region, is quantitatively evaluated. In Fig. 2.14, within the moveable region of the robot arm is enclosed by a dotted line in working coordinates, when the robot arm is moved along the arrow direction from each beginning point (p_x^0, p_y^0) (bullet • in figure) of 188 points divided in each 0.2[m], the value of $(\varepsilon_x, \varepsilon_y)$ about position of working linearized approximation error is calculated by (2.53) (line from bullet • in figure) and its results are illustrated. The length of the arm is $l_1 = 0.7$[m], $l_2 = 0.9$[m]. The motion velocity is $v_x = 0.1$[m/s], $v_y = 0.1$[m/s]. The symbol of inverse kinematics (2.39b) of the robot arm is often positive. From Fig. 2.14, the approximation precision of the working linearized model deteriorates near the boundary of the moveable region along the motion direction of the robot arm. Moreover, in the shrinking region of the robot arm, the working linearized approximation error becomes large. Since the working linearized approximation error is dependent on the posture of the arm but absolutely independent on the position in working coordinates of the arm, the results of the working linearized approximation error in Fig. 2.14 expresses that, the robot arm is moved not only along the error direction, but also rotated around the original point in Fig. 2.14 along any direction, and also the movement direction of arm is along

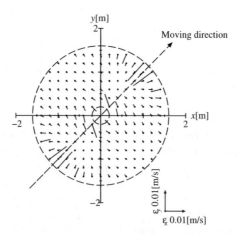

Fig. 2.14. Working linearized approximation error for various initial position (bullet •: initial position of robot arm; division from bullet •: working linearized approximation error vector $(\varepsilon_x, \varepsilon_y)$)

2.3 Linear Model of the Working Coordinates of an Articulated Robot Arm   45

the arrow direction in the figure and it is the dependent item of the working linearized approximation error.

Next, when changing the view point, from one beginning point of the robot arm (the distance from the initial point to the arm tip position is written as $r = \sqrt{(p_x^0)^2 + (p_y^0)^2}$), how the working linearized approximation error changes along various motion directions can be seen. At four points $r = 0.25, 0.38, 1.5, 1.55$[m] and with motion velocity $v = \sqrt{v_x^2 + v_y^2} = \sqrt{0.02} \approx 0.141$[m/s], when the arm is moved one cycle 2π at each direction with regarding initial position as the center, the results of position dependent item size $\sqrt{\varepsilon_x^2 + \varepsilon_y^2}$ of the working linearized approximation error are illustrated in Fig. 2.15. The horizontal axis ϕ of Fig. 2.15 represents the movement angle of arm. From the angle standard $\phi = 0$[rad] of angle stretching direction, $\phi = \pi$[rad] denotes the arm shrinking direction. From Fig. 2.15, at $r = 0.25$[m] and 1.55[m] near the boundary of the arm moveable region ($0.2 \leq r \leq 1.6$[m]), the working linearized approximation error becomes large at the arm stretching action. In the movement at the pull-push direction and vertical direction, the working linearized approximation error becomes fairly small.

When the working linearized approximation error (2.54) is dependent on time, the time shift with $K_p = 15$[1/s] of the time depending item $\lambda(t)$, is illustrated in Fig. 2.16. In the reference input time interval $\Delta T = 0.02$[s], $\lambda(t)$ is 0.0027[s]. From Fig. 2.15, the position dependent item size $\sqrt{\varepsilon_x^2 + \varepsilon_y^2}$ of the working linearized approximation error is below 0.001[m/s] with any direction motion within the region $0.38 \leq r \leq 1.5$[m]. Therefore, the maximum of the working linearized approximation error is 0.0027[mm]. This value is about 0.1% of the small interval length 0.141[m/s] $\times 0.02$[s] $= 0.00282$[m] with reference input time interval $\Delta T = 0.02$[s] and it is very small value. That is, when the reference input time interval is 0.02[s] with the robot arm motion velocity 0.141[m/s], the working linearized approximation error is within 0.1%

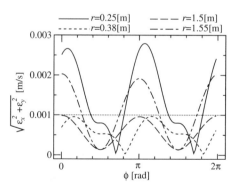

Fig. 2.15. Working linearized approximation error for various movement direction ϕ, initial position of robot arm r ($r = 0.25$[m], $r = 0.38$[m], $r = 1.5$[m], $r = 1.55$[m])

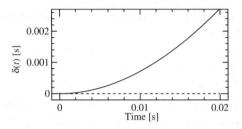

Fig. 2.16. Time dependence of working linearized approximation error $\lambda(t)$

of the objective trajectory in one division scale of objective trajectory, and the working linearizable approximation possible region can be as $0.38 \leq r \leq 1.5[\text{m}]$.

To a general robot arm, the derivation procedure of the working linearizable region is arranged. The length of link of the robot arm is l_1, l_2. The position loop gain is K_p. The reference input time interval is ΔT. The arm motion velocity is v. The distance from the initial point to the arm tip position is r (without losing generality, robot arm tip is on the x axis). The size of the working linearized approximation error at motion direction ϕ can be calculated by the following method.

1. Set $(p_x^0, p_y^0) = (r, 0)$, $(p_x^{\Delta T}, p_y^{\Delta T}) = (p_x^0 + v\Delta T \cos\phi, p_y^0 + v\Delta T \sin\phi)$
2. Using inverse kinematics (2.39), (θ_1^0, θ_2^0), $(\theta_1^{\Delta T}, \theta_2^{\Delta T})$ can be worked out.
3. The motion velocity in working coordinate velocity $(v_x, v_y) = (v \cos\phi, v \sin\phi)$ and the motion velocity in joint coordinate $(v_1, v_2) = ((\theta_1^{\Delta T} - \theta_1^0)/\Delta T, (\theta_2^{\Delta T} - \theta_2^0)/\Delta T)$ are calculated.
4. Using equation (2.53), the position dependent item of the working linearized approximation error $(\varepsilon_x, \varepsilon_y)$ is calculated. And its size $\sqrt{\varepsilon_x^2 + \varepsilon_y^2}$ is calculated.
5. Using equation (2.44), the time dependent item of the working linearized approximation error $\lambda(\Delta T)$ is calculated.
6. The size of working linearized approximation error $\sqrt{\varepsilon_x^2 + \varepsilon_y^2} \lambda(\Delta T)$ is calculated.

The working linearizable approximation possible region is defined with the region in which the working linearized approximation error for one interval of objective is below $\xi\%$ of small interval $v\Delta T$ divided of objective trajectory. In the working linearizable approximation possible region, the distance r from one initial point to the arm tip position is changed and the size of the working linearized approximation error $\sqrt{\varepsilon_x^2 + \varepsilon_y^2} \lambda(\Delta T)$ along two arm motion direction $\phi = 0 \sim 2\pi$ is calculated. Its size can be judged whether or not it is below the allowance $\xi v \Delta T / 100$.

2.3 Linear Model of the Working Coordinates of an Articulated Robot Arm 47

(3) Accumulation of Errors in the Working Linearized Model

In the evaluation so far, the appropriation of linearized approximation for divided one scale is evaluated on objective trajectory division. The derivation method of the working linearizable approximation possible region is given. The actual objective trajectory corresponds with the division trajectory and the working linearized approximation error of each region in the whole trajectory is checked about the time duration and how to integral.

(i) Shift of the working linearized approximation error in one region

In 2.3.1(2) and 2.3.1(3), for making objective trajectory at the beginning point of divided small interval, actual trajectory and position of the working linearized approximation trajectory similar, the derivation of the actual trajectory and the position of the working linearized approximation trajectory is carried out. In this part, when there are different values among the objective trajectory at the beginning point, actual trajectory and the working linearized approximation trajectory respectively, the working linearized approximation error in one region is analyzed. The positions of objective trajectory at the initial moment in working coordinate and joint coordinate are respectively (u_x^0, u_y^0), (u_1^0, u_2^0). The actual trajectory in working coordinates and joint coordinates are respectively (p_x^0, p_y^0), (θ_1^0, θ_2^0). The position of the working linearized approximation trajectory in working coordinate is $(\hat{p}_x^0, \hat{p}_y^0)$.

When we put equation (2.42) into (2.40), solve $(\theta_1(t), \theta_2(t))$ by using initial condition (θ_1^0, θ_2^0) and put this solution into equations (2.38a), (2.38b), the actual trajectory of the robot arm is as

$$p_x(t) = l_1\cos\{\theta_1^0 + (u_1^0 - \theta_1^0)\sigma(t) + v_1\lambda(t)\}$$
$$+ l_2\cos\{\theta_1^0 + \theta_2^0 + (u_1^0 + u_2^0 - \theta_1^0 - \theta_2^0)\sigma(t) + (v_1 + v_2)\lambda(t)\} \quad (2.55a)$$
$$p_y(t) = l_1\sin\{\theta_1^0 + (u_1^0 - \theta_1^0)\sigma(t) + v_1\lambda(t)\}$$
$$+ l_2\sin\{\theta_1^0 + \theta_2^0 + (u_1^0 + u_2^0 - \theta_1^0 - \theta_2^0)\sigma(t) + (v_1 + v_2)\lambda(t)\} \quad (2.55b)$$

where

$$\sigma(t) = 1 - e^{-K_p t}. \quad (2.56)$$

When we put equation (2.47) into (2.46) and solve $(\hat{p}_x(t), \hat{p}_y(t))$ by using the initial condition $(\hat{p}_x^0, \hat{p}_y^0)$, the working linearized approximation trajectory is calculated by

$$\hat{p}_x(t) = \hat{p}_x^0 + (u_x^0 - \hat{p}_x^0)\sigma(t) + v_x\lambda(t) \quad (2.57a)$$
$$\hat{p}_y(t) = \hat{p}_y^0 + (u_y^0 - \hat{p}_y^0)\sigma(t) + v_y\lambda(t). \quad (2.57b)$$

From the error between equation (2.57) and (2.55), the error between the actual trajectory and the working linearized approximation trajectory can be calculated by

$$e_x(t) = \hat{p}_x(t) - p_x(t)$$
$$= \hat{p}_x(t) - \tilde{p}_x(t) + o\{\sigma(t)\}$$
$$= (\hat{p}_x^0 - p_x^0)e^{-K_p t} + \varepsilon_x \lambda(t) + \{(u_{x^0} - p_x^0) + p_y^0(u_1^0 - \theta_1^0)$$
$$+ l_2 \sin(\theta_1^0 + \theta_2^0)(u_2^0 - \theta_2^0)\}\sigma(t) + o\{\sigma(t)\} \qquad (2.58a)$$
$$e_y(t) = \hat{p}_y(t) - p_y(t)$$
$$= \hat{p}_y(t) - \tilde{p}_y(t) + o\{\sigma(t)\}$$
$$= (\hat{p}_y^0 - p_y^0)e^{-K_p t} + \varepsilon_y \lambda(t) + \{(u_{y^0} - p_y^0) + p_x^0(u_1^0 - \theta_1^0)$$
$$- l_2 \cos(\theta_1^0 + \theta_2^0)(u_2^0 - \theta_2^0)\}\sigma(t) + o\{\sigma(t)\}. \qquad (2.58b)$$

The $(\tilde{p}_x(t), \tilde{p}_y(t))$ transformation is the Taylor expansion one order approximation of the actual trajectory (2.55) as

$$\tilde{p}_x(t) = p_x^0 + l_1 \sin\theta_1^0 \{(u_1^0 - \theta_1^0)\sigma(t) + v_1\lambda(t)\}$$
$$+ l_2 \sin(\theta_1^0 + \theta_2^0)\{(u_1^0 + u_2^0 - \theta_1^0 - \theta_2^0)\sigma(t) + (v_1 + v_2)\lambda(t)\} \qquad (2.59a)$$
$$\tilde{p}_y(t) = p_y^0 - l_1 \cos\theta_1^0 \{(u_1^0 - \theta_1^0)\sigma(t) + v_1\lambda(t)\}$$
$$- l_2 \cos(\theta_1^0 + \theta_2^0)\{(u_1^0 + u_2^0 - \theta_1^0 - \theta_2^0)\sigma(t) + (v_1 + v_2)\lambda(t)\}. \qquad (2.59b)$$

The first item in equation (2.58) is the item based on the difference between the actual trajectory (p_x^0, p_y^0) at the initial time and the working linearized approximation trajectory $(\hat{p}_x^0, \hat{p}_y^0)$. The second item is the working linearized approximation error (2.54) derived when the objective trajectory in 2.3.2(1) is identical with the actual trajectory and the working linearized approximation trajectory. The third item is the error item based on the position difference between the objective trajectory and the actual trajectory at the initial moment. The fourth item is the error item according to the Taylor expansion one order approximation of equation (2.59).

At the initial time, the working linearized approximation error $(\hat{p}_x^0 - p_x^0, \hat{p}_y^0 - p_y^0)$ is deteriorated with an index along time from the 1st item in the final equation in (2.58).

(ii) Accumulation of the working linearized approximation error

From the previous discussion, the time shift characteristic of the working linearized approximation error is similar with the x axis and y axis. Therefore, the x axis is discussed here. The upper boundary of the working linearized approximation error is analyzed by using triangle inequality when the reference input time interval ΔT increases. The size of the working linearized approximation error in reference input time interval ΔT is restrained from equation (2.58) as

$$|e_x(\Delta T)| = |(\hat{p}_x^0 - p_x^0)e^{-K_p \Delta T} + \varepsilon_x \lambda(\Delta T)$$
$$+ \{(u_{x^0} - p_x^0) + p_y^0(u_1^0 - \theta_1^0) + l_2 \sin(\theta_1^0 + \theta_2^0)(u_2^0 - \theta_2^0)\}\sigma(\Delta T)$$
$$+ o\{\sigma(\Delta T)\}|$$
$$\leq E_0 e^{-K_p \Delta T} + E_1 \sigma(\Delta T). \qquad (2.60)$$

2.3 Linear Model of the Working Coordinates of an Articulated Robot Arm

$E_0 = |\hat{p}_x^0 - p_x^0|$, E_1 are the positive constants for expressing the size of the working linearized approximation error generated newly in one region. $\lambda(\Delta T) = o\{\sigma(\Delta T)\}$ is adopted in the transformation of the final equation. Similarly, the size of the working linearized approximation error in the N_{th} division of the objective trajectory can be restrained.

$$|e_x(N\Delta T)| \leq |e_x\{(N-1)\Delta T\}|e^{-K_p\Delta T} + E_N\sigma(\Delta T). \tag{2.61}$$

According to using (2.61) step by step, the upper boundary of the size of the working linearized approximation error in the Nth region is expressed as below based on the accumulation of the working linearized approximation error from initial value.

$$\begin{aligned}|e_x(N\Delta T)| &\leq |e_x\{(N-1)\Delta T\}|e^{-K_p\Delta T} + E_N\sigma(\Delta T) \\ &\leq [|e_x\{(N-2)\Delta T\}|e^{-K_p\Delta T} + E_{N-1}\sigma(\Delta T)]e^{-K_p\Delta T} + E_N\sigma(\Delta T) \\ &\leq E_0 e^{-NK_p\Delta T} + E_1\sigma(\Delta T)e^{-(N-1)K_p\Delta T} + \cdots + E_N\sigma(\Delta T) \\ &\leq E_0 e^{-NK_p\Delta T} + E_{\max}\sigma(\Delta T)(e^{-(N-1)K_p\Delta T} + e^{-(N-2)K_p\Delta T} + \cdots + 1) \\ &= E_0 e^{-NK_p\Delta T} + E_{\max}\sigma(\Delta T)\frac{1-e^{-NK_p\Delta T}}{1-e^{-K_p\Delta T}} \\ &= E_0 e^{-NK_p\Delta T} + E_{\max}(1-e^{-NK_p\Delta T}) \end{aligned} \tag{2.62}$$

where $E_{\max} = \max(E_1, E_2, \cdots, E_N)$ and using value $\sigma(\Delta T)$ of equation (2.56) in the derivation procedure. The first item of equation (2.62) represents the effect of the working linearized approximation error in the initial moment. The second item represents the accumulation value of the working linearized approximation error generated in each division region of the objective trajectory. From equation (2.62), even the division number N is big, the working linearized approximation error is not divergent and it converges to a limited determined value. This $E_{\max}$ in 2.3.2(1) is the constant value according to the error based on the differences when the working linearized approximation error equation (2.54) is derived by the objective trajectory, actual trajectory and working linearized approximation trajectory. Moreover, when the dynamics of robot arm is good, i.e., the position loop gain K_p is big, the upper boundary of integral of the working linearized approximation error is $E_{\max}$. In addition, when the time $N\Delta T$ of the objective trajectory is constant, the division number of the objective trajectory is big and the division time is short, i.e., $N \to \infty$, $\Delta T \to 0$, the upper boundary of the integral value of the working linearized approximation error is unchanged in (2.62). If we evaluate this upper boundary of error, when the robot arm is moved from $(-0.8, 0.8[m])$ to $(-0.7, 0.8[m])$ with $0.1[m/s]$ at the positive direction of x axis under the same conditions with 2.3.2(2), the actual working linearized approximation error at the end point is 6.89×10^{-3}[mm]. From equation (2.62), the upper boundary of the calculated error is 6.10×10^{-2}[mm]. The working linearized approximation error accumulated by the error upper boundary is of such a size that it can be neglected.

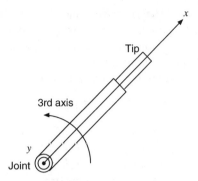

Fig. 2.17. Three-degree-of-freedom robot arm x axis and the third axis

(4) Expansion to a Multi-Axis Robot Arm

In the discussion so far, the working linearizable region for a two-axis robot arm is derived. In this part, the working linearizable region is discussed from a two-axis robot arm to a multi-axis robot arm. Concerning the SCALAR robot whose the third axis is the direct movement along z axis, the working linearizable region is to move the working linearizable region of a two-axis robot arm along the z axis direction. Since the 4th axis is the self-rotation of the end-effect, there is no need to make operation linearizable approximation.

Next, when determining the position in the working coordinates of a six-axis robot arm, three axes are considered from the base. The third axis of this six-axis robot arm is adopted as the y axis of a two-axis robot arm expressed by Fig. 2.11 for rotation. Therefore, to these three axes, the moveable region of robot arm is a ball with an empty center hole. The robot arm at the plate made by the third axis and x axis is illustrated by Fig. 2.17. When making a linearizable approximation in this plate by this third axis and x axis, it is different from the linear approximation of a two-axis robot arm discussed in the previous part. The former is that one axis is rotated and one axis is moved directly. The latter is that both axes are rotated. That is to say, the linear approximation of the two-axis robot arm discussed in the previous part means that the transformation from two axes rotation to two axes direct movement is possible. Therefore, the transformation from one axis rotation and one axis direct movement in the formed plate by the third axis and x axis is same as the linear approximation discussion of the two-axis robot arm discussed in the former part. That is to say, the robot arm in the formed plate by the third axis and x axis is possibly linearly approximated in working coordinates. The working linearizable region of the three-axis robot arm is the region that the working linearizable region of the two-axis robot arm is rotated by the y axis. Considering the third axis similar to hand, it is no need to make the operation linear approximation for self-rotation of the end-effect at the ball surface regarding the hand tip as the center in the operation space.

2.3.3 Adaptable Region of the Working Linearized Model and Experiment Verification

In order to observe the operation linearized approximation about control performance of the robot arm discussed so far, a computer simulation is carried out. The robot arm for simulation is $l_1 = 0.7[m]$, $l_2 = 0.9[m]$, $K_p = 15[1/s]$. The objective trajectory is to move $0.15[m]$ in the direction of the y axis with a velocity $0.25[m/s]$ and then to move $0.15[m]$ in the direction of the x axis. In the objective trajectory, the working linearized approximation error is within 0.2% and the working linearizable region is $(p_x^0, p_y^0) = (-0.8, 0.65[m])$ within $0.5 \leq r \leq 1.45[m]$ and $(p_x^0, p_y^0) = (-1.13137, 0.98137[m])$ out of possible region. Then the simulation is carried out. The reference input time interval is $\Delta T = 20[ms]$.

The operational linear approximation error in the top point $(x, y) = (-0.8, 0.8[m])$ within the working linearizable region is $(\varepsilon_x, \varepsilon_y) = (0.68, -0.09[mm/s])$. The working linearized approximation error $0.0018[mm]$ generated in one region of the objective trajectory is 0.037% of the divided objective trajectory $5[mm]$ and therefore it is very small. The working linearized approximation error in the top point $(x, y) = (-1.13137, 1.13137[m])$ out of the working linearizable region is $(\varepsilon_x, \varepsilon_y) = (0.0, 25.0[mm/s])$. The working linearized approximation error $0.675[mm]$ is generated within one region of the objective trajectory is 13.5% of the divided objective trajectory $5[mm]$.

In Fig. 2.18, the comparison of (a) response locus of linear approximated actual locus in the working linearizable approximation possible region and (b) the response locus of a linear approximated actual locus out of the working linearizable region about the two-axis robot arm is shown. In the working linearizable region of Fig.(a), the response locus of linear approximated is consistent with the actual locus in the figure. The maximal error of them is

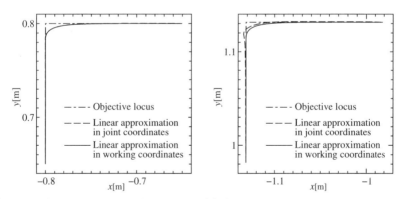

(a) Inside of working linearizable region (b) Outside of working linearizable region

Fig. 2.18. Comparison between linear approximation in joint coordinates and in working coordinates for a two-degree-of-freedom robot arm

0.2[mm], which can be neglected. Out of the working linearizable approximation possible region of Fig.(b), the response locus linear approximated has deviation with the actual locus. The maximal error of them is 2.7[mm], which is quite large. Moreover, out of the working linearizable region, overshoot is generated in the actual trajectory and the control performance of the robot arm itself is degraded. Besides, in the working linearizable region, when obtaining near equivalence between the actual trajectory of robot arm and the working linearized approximation trajectory, the control performance of the robot arm can be evaluated in working coordinates. However, out of the working linearizable region, the evaluation of the robot arm in working coordinates becomes difficult and control performance also deteriorates from the control performance expressed by the working linearized model.

Next, for illustrating the appropriation of the linear approximated model, the contour control experiment on a six-axis industrial robot arm (Performer K3S, maximal load is 3[kg]) was carried out (refer to experimental device E.2). The experimental results are shown in Fig. 2.19. The experimental results are almost the same as the simulation results in the working linearized model in Fig. 2.18(a). From this point of view, in the working linearizable region derived in this section, the working linearized model expressing the industrial robot arm can be verified by experiment.

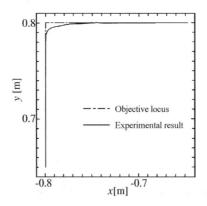

Fig. 2.19. Experimental results in the working linearizable region of the six-degree-of-freedom robot

3

Discrete Time Interval of a Mechatronic Servo System

The servo controller of a mechatronic system consists of the reference input generator, the position control part, the velocity control part, the current control part and the power amplifier part. By this controller, the motor is rotated and the mechanism part connected with the motor is moved. 15 years ago, the servo controllers were almost all constructed in hardware. In recent years, the reference input generator, position control part and velocity control part are digitally implemented using a micro processor and the current control part is analogically implemented. When the micro processor is installed into the closed-loop of the control system, this system must be considered as the sampling control system.

In this chapter, this sampling control system is different from the general discrete system. With the prerequisite that the dead time is very long, the relationship between the sampling time interval and contour control precision in the position loop and velocity loop, and also the relationship between the time interval of the command generation and the locus irregularity generated in the contour control as well as velocity fluctuation are discussed.

3.1 Sampling Time Interval

In the sampling control of the position loop and the velocity loop in the contour control of the mechatronic servo system, for calculating the control input in the next sampling period when the state has been known, the dead time is equivalent to the sampling time interval should be explained. Moreover, for making the control input as the 0th order hold, the constant control input should be the constant within the sampling time interval and there should be a big dead time for the entire system. According to experience, the sampling frequency, for the desired control performance which has no overshoot of locus in the contour control, is needed to be a value that is more than 30 times that of the entire cut-off frequency of the mechatronic servo system. However, there is no quantity analysis.

The mechatronic servo system is expressed by the 1st order system. In order to generate no oscillation (overshoot condition) in this transient response, the dead time equivalent to several sampling time interval was introduced. In addition, its cut-off frequency is not smaller than the cut-off frequency of the system without including dead time. By calculating the sampling frequency which satisfies the above two conditions, the relation of equation (3.6) $f_s \geq 27.5 f_{c1}$ can be derived.

By using the obtained equation, the proper sampling frequency in the sampling control system can be determined. It means that, it not only can prevent any decrease of the control performance of the mechatronic servo system generated with the low sampling frequency, but also can save the waste of the sampling control of high sampling frequency over the necessity. Moreover, in order to declare theoretically the reason for deterioration of the contour control performance with the rough sampling time interval including the dead time of the computing time, if there is dead time compensation in the control strategy, the control performance can be satisfied even with the rough sampling time interval.

3.1.1 Conditions Required in the Mechatronic Servo System

In the control of a mechatronic servo system, such as a robot arm, table of machine tool, etc, there are many kinds of sampling control using computers. When performing the contour control of a robot arm or machine tool, it is extremely important to avoid the overshoot of objective value (refer to 1.1.2 item 3). However, when this sampling control of the servo system is performed using a low sampling frequency sampler, the state measurement, control input calculation as well as the control signal output needs at least one sampling time interval. If it is dead time, there will appear an overshoot or oscillation in the output and also a deterioration of control performance according to general experience. The control law for compensating for dead time is actively studied theoretically [15]. But this kind of compensation method is with complicated control law. It cannot be adopted generally in the actual industrial servo system control. Therefore, in order to not generate control deterioration without performing dead time compensation, the sampler with a high sampling frequency is adopted and from one to several [kHz] frequencies is adopted for safety in the current industrial robot. If the sampler of high sampling frequency is adopted in the unnecessary case in the sampling control, the cost of hardware will be over the necessary expense for realizing a sampler of high frequency.

For the calculation of the control input in the velocity command without whole time. The transfer function of the 1st order system of the desired state without delay when output the control input obtained from the observed a value is written as (refer to item 2.2.3)

$$G_1(s) = \frac{K_p}{s + K_p} \tag{3.1}$$

where K_p denotes K_{p1} of equation (2.23) in the low speed 1st order model of item 2.2.3. The cut-off frequency of this servo system is $f_{c1} = K_p/2\pi$. For only including the delay frequency factors from the cut-off frequency, the possibility that can be of tracing correctly objective of this servo system should be hold in relation with the smooth objective trajectory. However, when performing the sampling control of this servo system and outputting the control input, the dead time actually exists due to the calculation delay of control input in the controller and the delay in reading states. For these cases, the servo system contains the sum L_1 of various dead times. When this sum of dead times is q_1 (q_1 is an integer over 1) times of the sampling time interval, there is $L_1 = q_1 \Delta t_p$ (Δt_p: sampling time interval). There has also the relation between the dead time and sampling frequency as $L_1 = q_1/f_s$ (f_s: sampling frequency).

If the sampling frequency of the sampling control is low, for this dead time, the overshoot and oscillation in the transient response occurred. The control performance has deteriorated. This overshoot is avoided completely in the contour control of the servo system (refer to the 1.1.2 item 3). For an understanding of the relation between control property of the servo system and the sampling frequency in the sampling control, the theoretical decision of the necessary sampling frequency for keeping control performance should be carried out. Therefore, in the sampling control, the dead time is only focused on and the effect of discretization is neglected. Based on this approximation, the strict analysis of the problem in the Z domain can be expressed in the s domain approximately. Hence, the following simple analysis can be carried out.

The transfer function of the 1st order system with dead time is as

$$G_{L1}(s) = \frac{K_p e^{-L_1 s}}{s + K_p e^{-L_1 s}}. \tag{3.2}$$

In this servo system with dead time, the conditions required from control properties are considered. In the servo system, the required control performance in the contour control is pursued correctly without overshoot for the complex objective trajectory with transient response of the servo system. Therefore, after arranging the required control performance, the two following conditions can be summarized.

(A) There is no divergence and no oscillation in the transient response (**overshoot condition**)

(B) The cut-off frequency of the system with dead time is not smaller than the cut-off frequency of the desired state (**cut-off frequency condition**)

The sampling frequency satisfying these two (A), (B) conditions simultaneously is calculated as below.

3.1.2 Relation between Control Properties and Sampling Frequency

(1) Relation Equation for the Overshoot Condition

The sampling frequency satisfying the overshoot condition of condition (A) imposed into the servo system is calculated.

In the transfer function (3.2) including dead time, by using the Pade approximation $e^{-L_1 s} \approx (2 - L_1 s)/(2 + L_1 s)$ of the dead time factors is easily adopted for analysis, the transfer function of equation (3.2) is approximately expressed as

$$G_{P1}(s) = \frac{K_p \left(\frac{2}{L_1} - s\right)}{s^2 + \left(\frac{2}{L_1} - K_p\right) s + \frac{2K_p}{L_1}}. \tag{3.3}$$

In order to satisfy the overshoot conditions that the servo system with dead time does not generate oscillations in the transient response and converge, the characteristic roots of equation (3.3) should be all negative. If this condition equation has several negative roots when the judgment equation of the characteristic equation is positive, the relation equation between the sampling frequency and cut-off frequency is obtained as

$$f_s \geq 18.3 q_1 f_{c1}. \tag{3.4}$$

However, in the transfer function of the Pade approximation of equation (3.3), which including unstable zero ($s = 2/L_1$), a few undershoots at the initial stage of the response are generated [16]. But the undershoots do not occur in the previous dead time system because the dead time is dealt with in the Pade approximation. The approximation error of the Pade approximation of dead time is bigger at the initial stage of response and tends to decrease with index function with time. The Pade approximation error in the delay time band in terms of overshoot possibly occurred according to the characteristic root is almost neglected. Therefore, the overshoot found in the approximated error is actually neglected. Only the overshoot in the characteristic root is discussed.

(2) Relation Equation for the Cut-Off Frequency Condition

The cut-off frequency condition of condition (B) is discussed here. Firstly, the cut-off frequency of the servo system of the desired state is $f_{c1} = K_p/2\pi$. On the other hand, the cut-off frequency of the servo system including dead time can be calculated by the following equation obtained from transfer function (3.3) by using Pade approximation.

$$f_{cP} = \frac{1}{2\pi} \left\{ \frac{1}{L_1} - \frac{K_p}{2} - \sqrt{\left(\frac{1}{L_1} - \frac{K_p}{2}\right)^2 - \frac{2K_p}{L_1}} \right\} \tag{3.5}$$

where, f_{cP} must be bigger than f_{c1} in order to satisfy the cut-off frequency condition. The condition, that f_{cP} is bigger than f_{c1}, can be held with the L_1 value when satisfying the overshoot condition (A).

3.1.3 Sampling Frequency Required in the Sampling Control

For a system with general dead time $q_1 \Delta t_p$, the relation equation (3.4) of the sampling frequency can be adopted in the sampling control problem of a servo system commonly existing the 0th order hold and dead time calculation of one sampling.

The continuous signal $f(t)$ is sampled in terms of the sampler (discretization). By the 0th order hold, the quantization error combining with the middle value of one sampling time interval is ignored. Therefore, for the previous signal $f(t)$, the delay with 1/2 sampling time can be found. In this sampling control, 1/2 sampler considering the 0th order hold and the generation of dead time in one sampling time from the calculation time is concerned. Hence, there are a total of 1.5 sampling time delays. The sum of the dead time is $L_1 = 1.5 \Delta t_p$. With $q_1 = 1.5$ in the relation equation (3.4) of the sampling frequency, it can be obtained that

$$f_s \geq 27.5 f_{c1}. \tag{3.6}$$

This result is almost equal to the value of sampling frequency known from experience, which is necessarily over 30 times that of the cut-off frequency.

According to the above, the experience value of about 30 times should be considered in theory.

3.1.4 Experimental Verification of the Sampling Frequency Determination Method

The servo system device used in the experiment consists of the table driven by a 0.85kW DC servo motor and ball spring, a servo controller (Yaskawa motor CPCR-MR-CA15) and a personal computer (NEC-PC9801). In the part of servo controller and the DC servo motor, the velocity loop is formed. Moreover, in the computer, the position loop is constructed. In this case, the velocity loop gain is $K_v = 185[1/s]$ and the position loop gain is $K_p = 1[1/s]$ as well as $K_v \gg K_p$. The part of velocity loop can be approximated by the direct connection (i.e. 1) in the block diagram. The overall servo system is expressed by the 1st order system of equation (3.1). If K_p is set with a small value, the remarkable deterioration in the sampling time interval can be illustrated. According to the signal flow, the position information of the DC servo motor can be obtained by integrating the tachogenerator signal read in the computer. The velocity command signal, calculated by the error of the position information and position command, is added into the servo controller through a D/A converter. Then, the velocity control is performed analogically

58   3 Discrete Time Interval of a Mechatronic Servo System

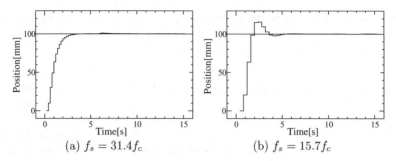

(a) $f_s = 31.4 f_c$ (b) $f_s = 15.7 f_c$

Fig. 3.1. Experimental results of the positioning control using shaft-driven device

by the DC servo motor according to the servo controller. Here, the sampling time interval is changed freely using the computer in the position loop. For verifying the effectiveness, this part is implemented by hardware as the digital (software) servo.

The experimental results are illustrated in Fig. 3.1. When satisfying the relation equation (3.6) of the necessary sampling frequency in $f_s = 31.4 f_{c1}$ of Fig.(a), there is the transient response wave which is almost equal to the simulation results of the desired state without dead time. Thus, the desired control properties can be obtained. When the relation equation (3.6) with rough sampling time interval is not satisfied for the $f_s = 15.7 f_{c1}$ of Fig.(b), the overshoot was generated in the transient response and the control properties was decreased. Besides, the amplitude of the stage variation of graph existed within the sampling time interval. From this point of view, the relation equation derived theoretically about the sampling frequency of equation (3.6) can be verified. In the experiment system with strict high order items, it is better to satisfy the relation equation (3.6) of the sampling frequency calculated with the 1st order approximation of the servo system.

3.2 Relation between Reference Input Time Interval and Velocity Fluctuation

In the servo controller, the general reference input generator is performed digitally. The objective trajectory generation needs computing time. The generated objective trajectory is then changed into the step-wise function (refer to 1.1.2 item 9) in a constant time interval (**reference input time interval**). From this discrete command signal, the **velocity fluctuation** of the reference input time interval in the performed servo system is generated.

In the current mechatronic system of the industrial field, for eliminating this velocity fluctuation, the position command of the step between the reference input time intervals is revised in the one-order hold value in each sampling time interval of the servo system. That is to say, the output between the reference input time interval is interpolated by line in each sampling time

3.2 Relation between Reference Input Time Interval and Velocity Fluctuation   59

interval. This is the method to produce consistency in the reference input time interval and the sampling time interval.

In this section, the theoretical relation equation (3.9) of the reference input time interval and the velocity fluctuation is derived. The steady-state velocity fluctuation is theoretically included when the strategy can be perfectly adopted based on the above industrial field pattern. Since the transient velocity fluctuation cannot be solved by the above method, the reason for velocity fluctuation generation is explained clearly.

For the servo controller in which the position loop is increased by hand, when the command of objective trajectory from outside based on the device on sale is given, the velocity fluctuation equivalent to equation (3.9) is generated because the conditions of the industrial field pattern is not satisfied. The occurred velocity fluctuation can be evaluated by the analysis results in this part.

3.2.1 Mathematical Model of a Mechatronic Servo System Concerning Reference Input Time Interval

(1) Velocity Fluctuation Generation within the Reference Input Time Interval

In the reference input generator of a mechatronic servo system, the objective position command values of each axis are calculated from the given operation task of the management part. At this time, the command values of an articulated robot should be transformed from working coordinates to joint coordinates. In addition, the curve part of the objective locus of an ellipse, etc, should be approximated by a line in the orthogonal type of NC machine tool. This necessary real calculation takes a long time, therefore, the reference input time interval is defined with a rough time interval. Thus, since the command input for the position control part is adopted when the objective value of each reference input time interval is given, sampled and held, the deviation of the rotational velocity of the motor, by the following velocity command part, current reference part and power amplifier part, and the velocity fluctuation in the response of the operation tip of the mechanism part driven by motor is also generated. Hence, the control performance deteriorates.

Generally, the velocity fluctuation factor of a mechatronic servo system often exists in the transient state and its variation is bigger than the steady state. Therefore, for avoiding the transition-state part and adopting a steady state, in fact, the utilization method for keeping the motion precision of the mechatronic servo system and the operational method for one axis are adopted. In this section, the velocity fluctuation of each reference input time interval as the study object cannot be avoided in the steady state of one axis operation. Moreover, since the velocity fluctuation factors of a mechatronic servo system are existed, it is very important to analyze them one by one. For this purpose, the analysis on the relationship between the reference input time interval of a mechatronic servo system and velocity fluctuation is more important than the analysis of their control performances.

(2) A Mathematical Model of a Mechatronic Servo System for Analyzing the Velocity Fluctuation

The model for analyzing the velocity fluctuation of each reference input time interval of a mechatronic servo system is constructed. The model of a mechatronic servo system for analyzing the relationship between the reference input time interval and velocity fluctuation can be expressed by the continuous 2nd order system illustrated in Fig. 3.2, where r denotes objective trajectory. ΔT denotes the reference input time interval in which the output command value from the reference input generator to the position control part. h_r denotes 0th order hold in the reference input generator. u_p denotes the position command value, K_p denotes the position loop gain, Δt_p denotes the sampling time interval in the position loop. h_p denotes the 0th order hold in the position control part. u_v denotes the velocity command value. K_v denotes the velocity loop gain. v denotes the velocity of motion. p denotes the position of motion. In the general operation, the objective trajectory r is the ramp input $r = v_{ref}t$ as the objective velocity v_{ref}.

The motion velocity of the mechatronic servo system of Fig. 3.2 is expressed as

$$\frac{dv(t)}{dt} = -K_v v(t) + K_v u_v(t) \tag{3.7}$$

where, K_v has the meaning of K_{v2} in the equation (2.29) of the middle speed 2nd order model in the 2.2.4 item. Moreover, k is the number of the reference input time interval. j is the sampling number of the position loop in ΔT ($0 \leq j < \Delta T/\Delta t_p$). The random moment can be expressed by $(k\Delta T + j\Delta t_p + t_p)$ ($0 \leq t_p < \Delta t_p$).

The position command value u_p is $u_p(k\Delta T + j\Delta t_p + t_p) = v_{ref}k\Delta T$ after sampling the objective trajectory $r(t) = v_{ref}t$ as the reference input time interval ΔT and making it with 0th order hold. Therefore, the velocity command value $u_v(k\Delta T + j\Delta t_p + t_p)$ can be expressed as

$$u_v(k\Delta T + j\Delta t_p + t_p) = (v_{ref}k\Delta T - p(k\Delta T + j\Delta t_p))K_p. \tag{3.8}$$

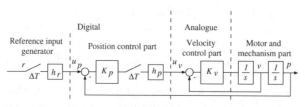

Fig. 3.2. 2nd order model of mechatronic servo system

3.2 Relation between Reference Input Time Interval and Velocity Fluctuation

When we input equation (3.8) into equation (3.7) and solving it on $v(t)$, then the motion velocity $v(k\Delta T + j\Delta t_p + t_p)$ can be as

$$v(k\Delta T + j\Delta t_p + t_p) = \left(1 - e^{-K_v t_p}\right)\left(v_{ref} k\Delta T - p\left(k\Delta T + j\Delta t_p\right)\right) K_p$$
$$+ v\left(k\Delta T + j\Delta t_p\right) e^{-K_v t_p}, \quad (0 \leq t_p < \Delta t_p). \tag{3.9}$$

The analytical solution can be easily found out.

The $e^{-K_v t_p}$ part of equation (3.9) expresses the change in Δt_p. The $v_{ref} k\Delta T - p(k\Delta T + j\Delta t_p)$ part expresses the change of Δt_p in ΔT. Based on them, the velocity fluctuation occurs in the mechatronic servo system illustrated in Fig. 3.2.

3.2.2 Industrial Field Strategy of the Velocity Fluctuation Generated in Reference Input Time Interval

(1) Equivalent Method in Sampling Time Interval to the Reference Input Time Interval

In the industrial field, this kind of velocity fluctuation should be avoided. One way is to let the reference input time interval ΔT be equal to the sampling time interval Δt_p of the position loop[18].

For the motion velocity equation (3.9) of the mechatronic servo system, when we input the condition $\Delta t_p = \Delta T$ and eliminate the initial value by adding the steady-state condition, the motion velocity can be as

$$v(k\Delta T + t_p) = \left(v_{ref} k\Delta T - p\left(k\Delta T\right)\right) K_p. \quad (0 \leq t_p < \Delta T) \tag{3.10}$$

The motion velocity $v(k\Delta T + t_p)$ is the constant within the reference input time interval ΔT. From this point of view, the steady-state velocity fluctuation of each reference input time interval does not occur.

Although this method is simple, the control performance has deteriorated because Δt_p is roughly adapted for ΔT and the position loop characteristic cannot be improved. In addition, since the reference input time interval must be reduced for shortening the ΔT to adapt for Δt_p, as a shortcoming, it is costly.

(2) Conversion Method of Each Reference Input Time Interval from the 0th Order Hold to the 1st Order Hold

Another method is that the 0th order hold h_r of each reference input time interval of each axis position command value calculated in the reference input generator is converted into the 1st order hold[17].

Since the ramp-shape objective trajectory $r(t) = v_{ref} t$ is 1st order hold with $v_{ref} t$ in ΔT, the position command is as $u_p(k\Delta t_p + t_p) = v_{ref} k\Delta t_p$, if $r(t) = v_{ref} t$ is 0th order hold in each sampling interval Δt_p of the position loop. Therefore, the motion velocity of equation (3.9) is changed as below after ΔT is replaced by Δt_p.

$$v(k\Delta t_p + t_p) = (v_{ref}k\Delta t_p - p(k\Delta t_p))K_p, \quad (0 \leq t_p < \Delta t_p). \qquad (3.11)$$

The motion velocity $v(k\Delta t_p + t_p)$ becomes constant within Δt_p. Therefore, the steady-state velocity fluctuation of each reference input time interval does not occur.

In this method, since ΔT can be lengthened and Δt_p can be shortened, it does not need to change the reference input time interval ΔT mostly into the short. However, for constructing the position control part by integer calculation, the 1st order hold of the position command value must be calculated into the integer value and then the fractional control is needed. The algorithm becomes complicated.

3.2.3 Parameter Relation between the Steady-State Velocity Fluctuation and the Mechatronic Servo System

(1) Velocity Fluctuation in the Steady State

The strategy of restraining the velocity fluctuation in the previous section can be adopted at any time without limitation. In recent years, a mechatronic servo system complete with the position loop has been on sale. The management part and the reference input generator are constructed by computer and therefore the simple mechatronic system can be constructed. By using this kind of product, it is difficult to adopt the strategy as introduced in the former section because the position loop is installed in advance. In recent years, the module robot and self-organized robot are studied widely. Since the axis number installed with the position loop and the number of robots are desired to be able to change freely. Moreover, there are many complex trajectory calculations when constructing the mechatronic servo system in the laboratory, adopting the strategy introduced in the former section is very difficult. Therefore, in this case, the theoretical analysis of the steady-state velocity fluctuation is important for the control performance prediction, design and adjustment.

For the mechatronic servo system with the states introduced as above, since $\Delta T \gg \Delta t_p$, the position loop can be continuously adopted. Therefore, the mathematical models of the position control part, velocity control part, motor part and mechanism part can be expressed as

$$\frac{d^2p(t)}{dt^2} = -K_v\frac{dp(t)}{dt} - K_vK_pp(t) + K_vK_pu_p(t) \qquad (3.12)$$

where K_p, K_v have the meaning of K_{p2}, K_{v2} in equation (2.29) of the middle speed 2nd order model in 2.2.4 item, respectively. Moreover, the input $u_p(t)$ is expressed by a step-wise function of

3.2 Relation between Reference Input Time Interval and Velocity Fluctuation

$$u_p(k\Delta T + t_p) = v_{ref} k \Delta T. \tag{3.13}$$

From equation (3.12) and (3.13), the velocity response of stage k in the steady state is as

$$v(k\Delta T + t_p) = \frac{p_1^s p_2^s}{p_2^s - p_1^s} \left(\frac{e^{p_2^s t_p}}{1 - e^{p_2^s \Delta T}} - \frac{e^{p_1^s t_p}}{1 - e^{p_1^s \Delta T}} \right) v_{ref} \Delta T \tag{3.14}$$

$$(0 \le t_p < \Delta T)$$

$$p_1^s = -\frac{K_v + \sqrt{K_v^2 - 4K_v K_p}}{2}$$

$$p_2^s = -\frac{K_v - \sqrt{K_v^2 - 4K_v K_p}}{2}.$$

Therefore, from the maximum value and minimum value in the reference input time interval ΔT of equation (3.14), the velocity fluctuation e_v^s can be calculated by

$$e_v^s = \frac{p_1^s p_2^s}{p_2^s - p_1^s} \left(\frac{1 - e^{p_1^s t_{max}^s}}{1 - e^{p_1^s \Delta T}} - \frac{1 - e^{p_2^s t_{max}^s}}{1 - e^{p_2^s \Delta T}} \right) v_{ref} \Delta T \tag{3.15}$$

$$t_{max}^s = \frac{1}{p_2^s - p_1^s} \log \frac{p_1^s \left(1 - e^{p_2^s \Delta T}\right)}{p_2^s \left(1 - e^{p_1^s \Delta T}\right)}. \tag{3.16}$$

For this purpose, the velocity fluctuation is generated in the steady state of a mechatronic servo system. Its size is proportional to the objective velocity v_{ref}.

(2) Application of the Analysis Results

By concerning the equation (3.15) expressing the relation between the reference input time interval ΔT derived in the last section and the velocity fluctuation e_v^s, the properties are obtained and graphed. Their application method is also discussed.

When graphing the property of the velocity fluctuation e_v^s, there are five related parameters: e_v^s, K_p, K_v, v_{ref}, ΔT. These parameters can be worked out by the relevant ratio of velocity fluctuation e_v^s/v_{ref}, gain ratio $n_{pv} = K_v/K_p$ and $K_p \Delta T$. When setting gain ratio $n_{pv} = K_v/K_p$ with 7, 10, 15, 20 respectively, Fig. 3.3 can be drawn with the vertical axis e_v^s/v_{ref} and horizontal axis $K_p \Delta T$. From this figure, the relevant ratio e_v^s/v_{ref} of velocity fluctuation is increased following the increase of $K_p \Delta T$ and gain ratio n_{pv}.

In the industrial field, the design procedures of a mechatronic servo system is that: firstly, the mechanism corresponding to the operational aim is designed and the properties of the constructed mechanism are tested; then the servo parameters (loop gains of velocity and position) without generating overshoot is determined from the tested property; finally, the digital controller which can implement the determined servo parameters is constructed.

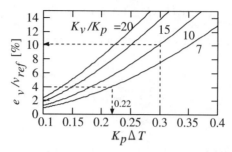

Fig. 3.3. Relative velocity fluctuation for $K_p \Delta T$

By using Fig. 3.3 and based on the velocity fluctuation, the controller design and machine type selection can be carried out. From the properties of the mechanism, when using gain $K_p = 20[1/\text{s}]$, $K_v = 140[1/\text{s}]$, the reference input time interval ΔT of the reference input generator is determined for making the relevant ratio of the velocity fluctuation to converge within $e_v^s/v_{ref} = 4[\%]$. From the figure of $n_{pv} = K_v/K_p = 7$ in Fig. 3.3 and the cross point of $e_v^s/v_{ref} = 4[\%]$, the $K_p \Delta T = 0.22$ can be read out. In order to make the reference input time interval below $\Delta T = (0.22/20) \times 1000 = 11[\text{ms}]$, the controller is designed or selected.

As another application method in Fig. 3.3, the parameters K_p, K_v, v_{ref}, ΔT of mechatronic servo system are given. The velocity fluctuation generation of this mechatronic servo system can be predicted beforehand. For example, if $K_p = 15[1/\text{s}]$, $K_v = 150[1/\text{s}]$, $v_{ref} = 50[\text{cm/s}]$, $\Delta T = 20[\text{ms}]$, $n_{pv} = K_v/K_p = 10$ is drawn in Fig. 3.3. From the cross point of $K_p \Delta T = 15 \times 0.02 = 0.3$, $e_v^s/v_{ref} = 10.0[\%]$ can be read out. Therefore, the velocity fluctuation is as $e_v^s = 50 \times 0.10 = 5.0[\text{cm/s}]$. The size of the generated velocity fluctuation in this mechatronic servo system can be known in advance.

3.2.4 Experimental Verification of the Steady-State Velocity Fluctuation

(1) Experimental Device and Experiment Conditions

In order to verify the property of the velocity fluctuation expressed by equation (3.15), the experiment using DEC-1 (refer to experiment device E.1) was carried out. The velocity loop gain of the servo controller of DEC-1 is $K_v = 100[1/\text{s}]$. The position loop gain is given as $K_p = 5[1/\text{s}]$ in the computer. The experiment was carried out with the reference input time interval $\Delta T = 40[\text{ms}]$, objective velocity $v_{ref} = 100[\text{rpm}]$, sampling time interval of the position loop $\Delta t_p = 1[\text{ms}]$ and control time 1[s]. The K_p is set with low value for remarkable variation.

(2) Experimental Result

The experimental results between 0.9~1 second with constant velocity fluctuation is illustrated in Fig. 3.4(a). The horizontal axis is time t[s], the vertical axis is velocity v[rpm] and the solid line is the velocity response. The velocity of motion is read in by computer the after A/D conversion of the tachogenerator. Since the high-frequent noise mixing, the remarkable velocity fluctuation occurred in each ΔT near the objective velocity 100[rpm]. For making comparison with this experimental results, the analysis velocity output based on the 2nd order model as equation (3.14) is shown by a broken line. From the figure, there are similar shapes of the wave from the experiment and the broken line. In the experiment, the mean of velocity fluctuation is 4.5[rpm]. It is almost same as the theoretical value $e_v^s = 4.3$[rpm] calculated from the equation (3.15) expressing the velocity fluctuation derived in the last section.

Next, the velocity fluctuation restraint strategy ($\Delta t_p = \Delta T$) illustrated in 3.2.2(1) was performed. The experimental results and simulation results of the mechatronic servo system are shown in Fig. 3.4(b). In this case, the velocity fluctuation with the reference input time interval occurred.

From the above experimental results, the equation (3.15) expressing the relation between reference input time interval which is derived by the model and the velocity fluctuation can be verified. This mathematical model as equation (3.7) is constructed based on the assumption, which is defined when constructing the continuous system mathematical model including the sampler derived for velocity fluctuation analysis of the mechatronic servo system.

The effectiveness of equation (3.15) was verified by the experiment of above one axis. Additionally, for a mechatronic servo system with an orthogonal motion, the expansion from one axis to multiple axes can be carried out. Since the articulated mechatronic servo system can be approximated in orthogonal coordinates (refer to section 2.3) in the possible region of linear approximation

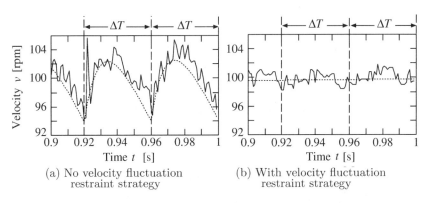

Fig. 3.4. Comparison between the experimental results using DEC-1 and the simulation results based on 2nd order model

within the general working region, the effectiveness of the proposed method can be also verified indirectly in the articulated mechatronic servo system.

3.2.5 Relation between Reference Input Time Interval and Transient Velocity Fluctuation

(1) Transient Velocity Fluctuation of the Mechatronic Servo System

In the industrial field, the controller of a mechatronic servo system which can restrain the velocity fluctuation is designed. In the mechatronic servo system which can restrain completely the steady-state velocity fluctuation, the hold circuit h_r between the reference input generator and position control part uses one-order hold circuit. The reference input time interval ΔT is set to be equal to the sampling time interval Δt_p of the position loop (refer to 3.2.2).

In this part, since the transient velocity fluctuation occurred even when restraining the steady-state velocity fluctuation, its analysis is carried out as below. As the control strategy, the transient velocity fluctuation when $\Delta T = \Delta t_p$ in 3.2.2(1) is adopted in the restraining the steady-state velocity fluctuation. In the continuous system, the mathematical model of the velocity control part, motor part and mechanism part is expressed as

$$\frac{dv(t)}{dt} = -K_v v(t) + K_v u_v(t). \tag{3.17}$$

If k is the stage of the reference input time interval ΔT, any moment can be expressed by $(k\Delta T + t_p)(0 \leq t_p < \Delta T)$. The position command value u_p is $u_p(k\Delta T + t_p) = v_{ref}(k+1)\Delta T$ by the 0th order hold when the objective trajectory $r(t) = v_{ref} t$ is sampled by the reference input time interval ΔT. Therefore, the velocity command value $u_v(k\Delta T + t_p)$ is expressed by

$$u_v(k\Delta T + t_p) = (v_{ref}(k+1)\Delta T - p(k\Delta T)) K_p. \tag{3.18}$$

When equation (3.18) is put into equation (3.17), by a inverse Laplace transform (refer to appendix A.1), the motion velocity $v(k\Delta T + t_p)$ is expressed as

$$v(k\Delta T + t_p) = \left(1 - e^{-K_v t_p}\right) (v_{ref}(k+1)\Delta T - p(k\Delta T)) K_p \\ + v(k\Delta T) e^{-K_v t_p}, \quad (0 \leq t_p < \Delta T). \tag{3.19}$$

Therefore, the analytical solution can be easily solved. This equation (3.19) is describing the damping of velocity command value changed stepwise within time constant $1/K_v$.

From the velocity of equation (3.19), in the zero infinite state (objective trajectory $r(t) = v_{ref} t$ is continuous) of the reference input time interval, the difference of velocity as

3.2 Relation between Reference Input Time Interval and Velocity Fluctuation

$$v_r(t) = v_{ref}\left\{1 + \frac{1}{p_1^s - p_2^s}\left(p_2^s e^{p_1^s t} - p_1^s e^{p_2^s t}\right)\right\} \quad (3.20)$$

$$p_1^s = -\frac{K_v + \sqrt{K_v^2 - 4K_v K_p}}{2}$$

$$p_2^s = -\frac{K_v - \sqrt{K_v^2 - 4K_v K_p}}{2}$$

is obtained with ΔT and using the maximum and maximum of maximal error (the first reference input time interval ($k = 1$) of the smallest damping), the maximal transient velocity fluctuation e_v^t is defined as

$$e_v^t = v(t_{\max}^t) - v_r(t_{\max}^t) \quad (3.21)$$

$$= v_{ref}\left[\Delta T K_p \left(1 - e^{-K_v t_{\max}^t}\right)\right.$$

$$\left. - \left\{1 + \frac{1}{p_1^s - p_2^s}\left(p_2^s e^{p_1^s t_{\max}^t} - p_1^s e^{p_2^s t_{\max}^t}\right)\right\}\right]. \quad (3.22)$$

However, $t_{\max}^t$ is calculated by

$$\Delta T e^{-K_v t_{\max}^t} + \frac{1}{p_1^s - p_2^s}\left(e^{p_2^s t_{\max}^t} - e^{p_1^s t_{\max}^t}\right) = 0. \quad (3.23)$$

(2) Graph of the Relationship Equation of the Transient Velocity Fluctuation

In the analytical solution equation (3.22), since using many parameters is difficult, the relation between frequently adopted parameters and the transient velocity fluctuation is graphed.

When $K_v = 100[1/s]$ is fixed, Fig. 3.5 illustrated the reference input time interval $\Delta T[s]$ when using $K_p = 1, 5, 10, 20[1/s]$ and the division $e_v^t/v_{ref}[\%]$ of the transient velocity fluctuation for the objective velocity. By using this figure, the relationship between the reference input time interval and the transient velocity fluctuation can be known.

3.2.6 Experimental Verification of the Transient Velocity Fluctuation

In order to verify the transient velocity fluctuation within the reference input time interval analyzed in the last part, an experiment was carried out using DEC-1(refer to experiment device E.1). The experimental conditions are $\Delta T = \Delta t_p = 40[\text{ms}]$, $K_p = 5[1/s]$ and the objective velocity $v_{ref} = 10.5[\text{rad/s}](100[\text{rpm}])$. The velocity response between 0.4 second from the beginning of control is shown in Fig. 3.6(a). Figure 3.6(b) shows the velocity fluctuation. Here, the horizontal axis is the time $t[s]$, the upper part of the vertical axis is the velocity $v(t)[\text{rad/s}]$ and the bottom part is the velocity

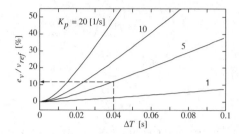

Fig. 3.5. Relation between velocity fluctuation e_v^s and reference input time interval ΔT ($K_v = 100[1/\text{s}]$)

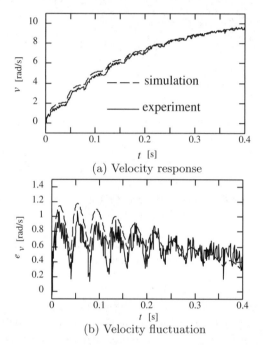

Fig. 3.6. Experimental results using DEC-1 and simulation results using 2nd order model

fluctuation $e_v^t(t)[\text{rad/s}]$. The solid line denotes the experimental result, and the dotted line is the simulation results analyzed strictly by using Neuman series for differential equation of (3.17) within 1[ms]. The characteristics of the transient velocity fluctuation between the experiment and the simulation are very close. In each reference input time interval $\Delta T = 40[\text{ms}]$, the velocity fluctuation occurred and then decreased slowly. In the experiment, the size of the initial maximal velocity fluctuation of the initial stage is 1.10[rad/s]. By using Fig. 3.5 for visualizing the equation (3.22), with $K_v = 100[1/\text{s}]$, $\Delta T = 40[\text{ms}]$ and $K_p = 5[1/\text{s}]$, the velocity fluctuation to objective velocity can be as $e_v^t/v_{ref} = 11.0[\%]$. Therefore, the theoretical value of the transient velocity fluctuation is $e_v^t = 0.110 \times 10.5 = 1.16[\text{rad/s}]$. It is almost the same as the experimental result. Based on the above, the effectiveness of the analysis results can be verified.

3.3 Relationship between Reference Input Time Interval and Locus Irregularity

The reference input time interval and the velocity fluctuation in the digital controller was introduced in the section 3.2. However, in the contour control, this fluctuation may occur on the surface of the product and this surface can be changed as rough expressed as locus irregularity. This locus irregularity may occur in each reference input time interval when the servo system property of each axis in the mechanism is not consistent. The generation mechanism of this locus irregularity and its quantitative analysis are expected.

The analytical solution of locus irregularity generated in each reference input time interval is given in equation (3.29).

By using the theoretical analysis solution of the locus irregularity, the prediction of movement precision of the robot or machine tool as well as the design arrangement of the mechatronic servo system of the required locus precision are possible.

3.3.1 Locus Irregularity in the Reference Input Time Interval

(1) Mathematical Model of the Orthogonal Two-Axis Mechatronic Servo System

For analyzing the relation between the reference input time interval of a mechatronic servo system and locus irregularity, firstly, the mathematical model of the orthogonal two-axis mechatronic servo system is constructed, and then its response in each reference input time interval is calculated. The relationship between the reference input time interval and the locus irregularity is analyzed quantitatively. Next, its analysis result is expanded into the joint coordinates and space coordinates. The general locus irregularity of the mechatronic system is discussed.

As the reason of deterioration of the control performance, the effect of coordinate transform and mechanism dynamics, the calculation time in the digital controller, the resolution of the encoder or D/A converter, cogging torque as well as stick-slip should be considered. Generally, when a mechatronic system is structured with multiple axes. But it is better to separately consider the problem of generation in each axis of servo system and the problem of generation of multi-axis structure (refer to 1.1.2 item 6).

The reference input generators and position control parts are always adopted with a digital controller. Since the position control part is simply used for computation, its computation cycle is carried out within the narrow sampling time interval. But the reference input generator performs complicated computation, such as inverse kinematics computation, etc. Therefore, its computation cycle is longer than the sampling time interval. According to this width of reference input time interval, the velocity fluctuation occurs at one axis and the locus irregularity occurs when combining two such axes. Therefore, the problem of the locus irregularity is firstly solved in the orthogonal two-axis mechatronic system with x axis and y axis, and then the problem of locus irregularity of the general mechatronic system with coordinate transform is solved.

With the general motion condition, the model of x axis and y axis in the orthogonal two-axis mechatronic servo system can be constructed with a 1st order system respectively (refer to the item 2.2.3)

$$\frac{dp_x(t)}{dt} = -K_{px}p_x(t) + K_{px}u_x(t) \qquad (3.24a)$$

$$\frac{dp_y(t)}{dt} = -K_{py}p_y(t) + K_{py}u_y(t) \qquad (3.24b)$$

where $p_x(t)$, $p_y(t)$ are positions in time t, $dp_x(t)/dt$, $dp_y(t)/dt$ are velocities, $u_x(t)$, $u_y(t)$ are servo system input of each axis, K_{px}, K_{py} have the meanings of K_{p1} in the low speed 1st order model equation (2.23) of item 2.2.3 at x axis and y axis

For a mechatronic system, in order to make the steady-state error values of each axis similar at the initial arrangement time of device, the position loop gain of the controller of each axis in servo system should be regulated. According to the motion condition and working load based on the arrangement, the property of the servo system will be changed slightly. There are existing the regulation error at the initial self-arrangement. Therefore, these summed errors accumulate the difference of position loop gain K_{px} of equation (3.24a), (3.24b) and K_{py} express the property of the mechatronic servo system with the 1st order system. The difference of K_{px} and K_{py} is the reason for the generation of locus irregularity.

(2) Response of a Mechatronic Servo System in Each Reference Input Time Interval

The locus irregularity, as the analysis object, occurred in the rough reference input time interval, occurred in the transient state with changeable input, cannot be found in the steady state. Generally, in the transient state, there have been other kinds of locus deterioration except this locus irregularity. Comparing with the transient state, the locus precision of contour control in the steady state can be improved. However, the locus irregularity in each reference input time interval in this section is the main reason of dominant rest contour control performance deterioration in the steady state. Wherein, the

3.3 Relationship between Reference Input Time Interval and Locus Irregularity

steady state analysis as the discussion point is performed. In the steady state, the response features with the reference input time interval is the transient response.

The aim of this analysis is to understand the quantitative relation between the reference input time interval and the steady state of locus irregularity. Therefore, the drawn objective locus of the mechatronic system is a straight line (the objective operation velocity of each axis is constant) and the input of the model of a mechatronic servo system as the equation $(3.24a)(3.24b)$ is constructed.

The objective working velocity of each axis is v_x, v_y, respectively. The input $u_x(t)$, $u_y(t)$ of each axis of the servo system calculated in each reference input time interval ΔT is expressed by the step-wise function of ΔT amplitude as

$$u_x(t) = v_x \Delta T U(t) + v_x \Delta T U(t - \Delta T)$$
$$+ v_x \Delta T U(t - 2\Delta T) + v_x \Delta T U(t - 3\Delta T) + \cdots \quad (3.25a)$$
$$u_y(t) = v_y \Delta T U(t) + v_y \Delta T U(t - \Delta T)$$
$$+ v_y \Delta T U(t - 2\Delta T) + v_y \Delta T U(t - 3\Delta T) + \cdots \quad (3.25b)$$

where $U(t)$ is the unit step function.

For analyzing the locus irregularity generated with a rough reference input time interval, the above equation $(3.24a)\sim(3.25b)$ are one of the main point of this analysis and their solutions can be easily obtained by the existed analysis method. Here, a Laplace transform (refer to the appendix A.1) is carried out in equation $(3.25a)$, $(3.25b)$, and put them into the equation $(3.24a)$, $(3.24b)$ which have been also transformed by a Laplace transform. Then the response in each ΔT can be solved. If performing an inverse Laplace transform (refer to appendix A.1), the response in one reference input time interval ΔT with big enough stage m of ΔT is as

$$p_x(m\Delta T + t) = v_x \Delta T \left(m - \frac{e^{-K_{px}t}}{1 - e^{-K_{px}\Delta T}} \right), (0 \leq t < \Delta T). \quad (3.26a)$$

$$p_y(m\Delta T + t) = v_y \Delta T \left(m - \frac{e^{-K_{py}t}}{1 - e^{-K_{py}\Delta T}} \right), (0 \leq t < \Delta T). \quad (3.26b)$$

For this purpose, since the input of the mechatronic servo system and the servo system can be clearly expressed by the equations $(3.24a)$, $(3.24b)$ and $(3.25a)$, $(3.25b)$, the response in each reference input time interval ΔT in the steady state can be clearly worked out. These response equations $(3.26a)$, $(3.26b)$ in each ΔT is adopted for the locus irregularity analysis in the next part.

(3) Theoretical Solution of the Locus Irregularity

From the response equation $(3.26a)$ and $(3.26b)$ in each reference input time interval ΔT, the time t is eliminated, and then the response locus of the

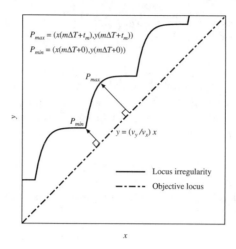

Fig. 3.7. Locus irregularity in mechatronic servo system

mechatronic system is obtained. The error between the locus of this mechatronic system and the objective locus is the **locus error** . This locus error is determined by the normal vector distance from the objective locus to the locus of the servo system. By the error of maximum value and minimum value of locus error in one reference input time interval, the locus irregularity is defined.

In Fig. 3.7, the response among many reference input time intervals of an orthogonal two-axis mechatronic servo system is shown. In Fig. 3.7, the horizontal axis is the x axis, vertical axis is the y axis and the dotted broken line is the objective locus $y = (v_y/v_x)x$. At the moment $(m\Delta T + t)$, the normal vector distance from objective locus $y = (v_y/v_x)x$ to locus coordinate $(x(m\Delta T + t), y(m\Delta T + t))$ is

$$e(t) = \frac{|v_y x(m\Delta T + t) - v_x y(m\Delta T + t)|}{\sqrt{v_x^2 + v_y^2}}. \tag{3.27}$$

When we put p_x, p_y of equation (3.26a), (3.26b) into x and y, the locus error $e(t)$ is as

$$e(t) = \frac{v_x v_y \Delta T}{\sqrt{v_x^2 + v_y^2}} \left| \frac{e^{-K_{px}t}}{1 - e^{-K_{px}\Delta T}} - \frac{e^{-K_{py}t}}{1 - e^{-K_{py}\Delta T}} \right|. \tag{3.28}$$

As shown in Fig. 3.7, if the locus is minimal position $P_{\min}$ at $t = 0$ and the maximal position $P_{\max}$ as $de(t)/dt = 0$, the locus irregularity e_m is as below by the error of maximal value and minimal value of the locus error $e(t)$ and using equation (3.28).

$$e_m = |e(t_m) - e(0)|$$

3.3 Relationship between Reference Input Time Interval and Locus Irregularity

$$= \frac{v_x v_y \Delta T}{\sqrt{v_x^2 + v_y^2}} \left| \frac{e^{-K_{px} t_m} - 1}{1 - e^{-K_{px} \Delta T}} - \frac{e^{-K_{py} t_m} - 1}{1 - e^{-K_{py} \Delta T}} \right| \quad (3.29)$$

where t_m is as below with $de(t)/dt = 0$

$$t_m = \frac{1}{K_{px} - K_{py}} \log \frac{K_{px}\left(1 - e^{-K_{py}\Delta T}\right)}{K_{py}\left(1 - e^{-K_{px}\Delta T}\right)}. \quad (3.30)$$

This equation (3.29) is the analytical solution of locus irregularity occurring in each reference input time interval ΔT. From equation (3.29), if the position loop gain K_{px} of the x axis and K_{py} of the y axis are the same, e_m is zero. In general, it is difficult to make the position loop gain K_{px} and K_{py} of the servo system in the mechatronic servo system absolutely the same, i.e., ($K_{px} \neq K_{py}$). As the reason, the generation of locus irregularity according to the equation (3.29) in each reference input time interval ΔT can be found from the above equation.

(4) Expansion to the Articulated Robot

The discussion on the analysis of locus irregularity occurred in the orthogonal two-axis mechatronic servo system, carried out at 3.3.1(3), is expanded to the articulated robot. The articulated robot with two axes is constructed with two rigid links and two joints, as illustrated in Fig. 2.11 of section 2.3. Each joint has a servo motor and is constructed by a position control system. Its each joint angle is controlled to follow the objective angle.

The mathematical model of each axis in the articulated robot shown in Fig. 2.11 is expressed as the following 1st order system with the same discussion with equation (3.24a) and (3.24b).

$$\frac{d\theta_1(t)}{dt} = -K_{p1}\theta_1(t) + K_{p1}u_1(t) \quad (3.31a)$$

$$\frac{d\theta_2(t)}{dt} = -K_{p2}\theta_2(t) + K_{p2}u_2(t) \quad (3.31b)$$

where $d\theta_1(t)/dt$, $d\theta_2(t)/dt$ are the angle velocities, K_{p1}, K_{p2} have the meanings of K_{p1} in the low speed 1st order model equation (2.23) of item 2.2.3 for each joint. $u_1(t)$, $u_2(t)$ are input of each axis.

For discussing the locus irregularity on the working coordinates (x, y) for this articulated robot, the relation with the locus irregularity in the joint coordinates (θ_1, θ_2) is worked out. The transformation from joint coordinates (θ_1, θ_2) to working coordinates (x, y) is expressed as (refer to section 2.3)

$$x = l_1 \cos(\theta_1) + l_2 \cos(\theta_1 + \theta_2) \quad (3.32a)$$
$$y = l_1 \sin(\theta_1) + l_2 \sin(\theta_1 + \theta_2). \quad (3.32b)$$

The transformation between two coordinates is a nonlinear transform. It adopts the linear transformation within the small part. The relation between

74   3 Discrete Time Interval of a Mechatronic Servo System

the slight change $(d\theta_1, d\theta_2)$ near (θ_1^0, θ_2^0) in the joint coordinates and the slight change (dx, dy) in the working coordinates is expressed by a one-order approximation of a Taylor expansion as

$$\begin{pmatrix} dx \\ dy \end{pmatrix} = J \begin{pmatrix} d\theta_1 \\ d\theta_2 \end{pmatrix} \qquad (3.33)$$

where J is the Jacobian matrix

$$J = \begin{pmatrix} -l_1 \sin(\theta_1^0) - l_2 \sin(\theta_1^0 + \theta_2^0) & -l_2 \sin(\theta_1^0 + \theta_2^0) \\ l_1 \cos(\theta_1^0) + l_2 \cos(\theta_1^0 + \theta_2^0) & l_2 \cos(\theta_1^0 + \theta_2^0) \end{pmatrix}. \qquad (3.34)$$

Moreover, by using the same Jacobian matrix J, two coordinates for velocity can be expressed as

$$\begin{pmatrix} \dfrac{dx}{dt} \\ \dfrac{dy}{dt} \end{pmatrix} = J \begin{pmatrix} \dfrac{d\theta_1}{dt} \\ \dfrac{d\theta_2}{dt} \end{pmatrix}. \qquad (3.35)$$

With the common motion condition, in the joint coordinates of the articulated robot, the model (3.31a), (3.31b) can be approximated by the model (3.24a), (3.24b) of an orthogonal two-axis mechatronic servo system (refer to section 2.3). In an articulated robot with the discussion of 3.3.1(1)~(3) by using (3.24a), (3.24b), the locus irregularity can be expressed approximately by the relation equation (3.29).

(5) Expansion to the Three-Axis Mechatronic Servo System

The discussion in 3.3.1(4) is the locus irregularity discussion on the plate of two axes. In this part, the locus irregularity discussion is expanded to three axes. In the expansion from two axes discussion to three axes, the z axis is added with the x axis and the y axis in the mechatronic servo system model (3.24a), (3.24b)

$$\frac{dp_z(t)}{dt} = -K_{pz}p_z(t) + K_{pz}u_z(t) \qquad (3.36)$$

where $p_z(t)$ is the position of the z axis, $dp_z(t)/dt$ is velocity, $u_z(t)$ is the input of servo system, K_{pz} has the meaning of K_{p1} in the low speed 1st order model (2.23) of item 2.2.3 in the z axis. The input $u_z(t)$ of servo system of the z axis is as

$$u_z(t) = v_z \Delta TU(t) + v_z \Delta TU(t - \Delta T)$$
$$+ v_z \Delta TU(t - 2\Delta T) + v_z \Delta TU(t - 3\Delta T) + \cdots. \qquad (3.37)$$

If calculating the response of the z axis after enough stage number m is put into equation (3.36), as similar as equation (3.26a), (3.26b), it can be obtained that

3.3 Relationship between Reference Input Time Interval and Locus Irregularity 75

$$p_z(m\Delta T + t) = v_z \Delta T \left(m - \frac{e^{-K_{pz}t}}{1 - e^{-K_{pz}\Delta T}} \right), \quad (0 \le t < \Delta T) \quad (3.38)$$

where v_z is the objective velocity of the z axis.

In the orthogonal plate with an objective locus, the locus error $e_3(t)$ is the distance with the space coordinates $(p_x(m\Delta T+t), p_y(m\Delta T+t), p_z(m\Delta T+t))$ of the servo system calculated according to the (3.26a), (3.26b), (3.38) about the objective space coordinates. By using the locus error at the moment of $t = 0$ and $de_3(t)/dt = 0$, the locus irregularity can be calculated by

$$e_{m3} = |e_3(t_{m3}) - e_3(0)| \quad (3.39)$$

where t_{m3} is the moment of $de_3(t)/dt = 0$.

Based on the above, the locus irregularity discussion about two axes can be expanded into the three axes.

3.3.2 Experimental Verification of the Locus Irregularity Generated in the Reference Input Time Interval

(1) Experimental Result of Locus Irregularity

For verifying the theoretical analysis results of equation (3.29) of locus irregularity in each reference input time interval derived in item 3.3.1, the experimental work was carried out using DEC-1 (refer to experiment deviceE.1). In a mechatronic system, since it is difficult to make the gain of the servo system of each axis exactly consistent, the locus irregularity occurs in each reference input time interval. This experiment imitates the actual situation. The DC servo motor is rotated two cycles by changing the conditions of one motor. The first rotation is the motion of the x axis and second rotation is the motion of the y axis. Combining the motion results of two rotations, the experiment of an orthogonal two-axis mechatronic servo system was carried out. The inconsistency of position loop gain of the servo system was realized by changing the setting of position loop gain K_p in the computer for experiment.

The control conditions are reference input time interval $\Delta T = 0.1[\text{s}]$, objective velocity $v_x = v_y = 6[\text{rad/s}]$, sampling time interval $\Delta t_p = 0.01[\text{s}]$, x axis ($K_p = 10[1/\text{s}] = K_{px}$) for the first rotation, y axis ($K_p = 11[1/\text{s}] = K_{py}$) for the second rotation. These control conditions are selected if the torque limitation (current limitation) of the servo driver need not be considered in the experiment.

The experimental results are shown in Fig. 3.8 and Fig. 3.9. Fig. 3.8 illustrates the objective locus and the results of the locus in the experiment of the orthogonal two-axis mechatronic servo system. The horizontal axis is the x axis position [rad]. The vertical axis is the y axis position [rad]. In Fig. 3.8, for checking the locus irregularity that occurred in experiment, the calculated locus error is given in Fig. 3.9. The horizontal axis is the motion distance [rad] combining the x axis and the y axis. The vertical axis is locus error [rad]. The

76   3 Discrete Time Interval of a Mechatronic Servo System

solid line is the experimental results and the dotted line is simulation results of the servo system using the 1st order system as (3.24a), (3.24b).

From Fig. 3.9, the steady-state error and occurred unevenness in each reference input time interval of the locus can be seen. Since this steady-state error is different from the error of considered object in this research, it is the reason of the response delay of control system. The unevenness generated in each reference input time interval is the locus irregularity which is the object of this research. This locus irregularity causes the coarseness of movement in the robot. From the figure, the locus irregularity is 4.44×10^{-3}[rad]. It is consistent with the calculated value 5.12×10^{-3}[rad] based on the theoretical analytical solution of equation (3.29). It proves that the theoretical analytical solution about locus irregularity is as almost same as the experimental results in terms of shape and values. Moreover, there are about 0.003[rad] difference in faces to 0.04[rad] in the steady-state error of locus error. However, from the overall point of view, the simulation is very consistent with the experiment.

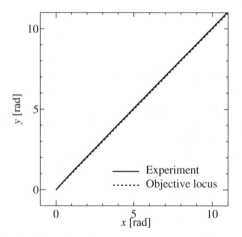

Fig. 3.8. Experimental result of locus irregularity

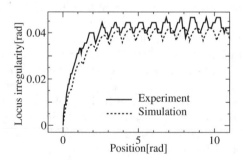

Fig. 3.9. Comparison between experimental results and simulation results based on 1st order model

3.3 Relationship between Reference Input Time Interval and Locus Irregularity   77

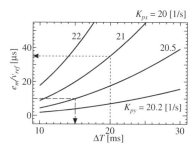

Fig. 3.10. Relation between reference input time interval and locus irregularity

Besides, the angle opening in the wave shape in the experimental results is caused by the rough encoder resolution 2000[pulse/rev].

Based on the above, the effectiveness of the relation equation (3.29) of the reference input time interval of the orthogonal two-axis mechatronic servo system in the steady state and locus irregularity was verified. According to the explanation in 3.3.1(4) and (5), it verified that the proposed method can be also adopted in the articulated robot because the experimental results can be approximated within allowance in the working linearizable region.

3.3.3 Application Value of the Theoretical Analysis Result

In this part, the application method of equation (3.29) of the theoretical analysis results of locus irregularity verified by experiment is discussed. From equation (3.29), with $0 \leq v_x, v_y \leq v_{ref}$, the size of locus irregularity becomes maximum when the objective operation velocity adopts $v_x = v_y = v_{ref}$ of the maximal value for both two axes. Therefore, if we put $v_x = v_y = v_{ref}$ into equation (3.29) as

$$e_m = \frac{v_{ref} \Delta T}{\sqrt{2}} \left| \frac{e^{-K_{px} t_m} - 1}{1 - e^{-K_{px} \Delta T}} - \frac{e^{-K_{py} t_m} - 1}{1 - e^{-K_{py} \Delta T}} \right| \quad (3.40)$$

the locus irregularity e_m is then proportional to the objective operation velocity v_{ref}.

This equation (3.40) is drawn in the graph. In order to understand the relationship of various parameters in a two-dimensional graph easily, the vertical axis is the locus irregularity e_m/v_{ref} for objective working velocity. The calculation results in the steady state about the objective working velocity in the reference input time interval ΔT is shown in Fig. 3.10. In the industrial field or a robot, there have been several percent to 10% difference amongst the gains of each axis of the servo system in the machine tool. For understanding the regions of these properties, the position loop gain of the x axis is fixed as $K_{px} = 20[1/\text{s}]$. The position loop gains of the y axis is changed as $K_{py} = 20.2$ (1[%]), 20.5 (2.5[%]), 21 (5[%]), 22 (10[%])[1/s] (% denotes the division of K_{px} and K_{py}). The locus irregularity is increased along the increment of the reference input time interval ΔT. In addition, the deviation of K_{px} and K_{py} can be easily found visually with their increment. By using this graph, if the deviation of K_{px} and K_{py} of the mechatronic servo system is known, the occurrence of locus irregularity can be predicted in advance. Concretely, the gains are $K_{px} = 20[1/s]$ and $K_{py} = 21[1/s](5\%$ error), the reference input time interval is $\Delta T = 20[\text{ms}]$ and the objective operation velocity is $v_{ref} = 0.4[\text{m/s}]$. As shown by the dotted line in Fig. 3.10, the locus irregularity e_m/v_{ref} of the objective velocity is $35[\mu s]$ for $\Delta T = 20[\text{ms}]$ and $K_{py} = 21[1/s]$. If the objective operation velocity is $v_{ref} = 0.4[\text{m/s}]$, the locus irregularity is then $0.4 \times 35 = 14[\mu m]$.

In general, there are many reasons for locus irregularity. For restraining it, the encoder resolution is always raised and the sampling time interval of the position loop is shortened in the industrial field. Based on the theoretical analysis, it is known the fact that the locus irregularity occurred in the reference input time interval ΔT is the main effect on the locus precision.

Next, by using the Fig. 3.10 graphing the analysis results, how many reference input time intervals determines the control precision is discussed. If the position loop gains of a mechatronic servo system are $K_{px} = 20[1/s]$ and $K_{py} = 20.5[1/s](2.5\%$ error), objective operation velocity is $v_{ref} = 0.1[\text{m/s}]$ and locus irregularity is below $1[\mu m]$, the reference input time interval ΔT can be worked out. Since the objective working velocity is $v_{ref} = 0.1[\text{m/s}]$, locus irregularity is $e_m/v_{ref} = 10[\mu s]$. From the broken line in Fig. 3.10, ΔT can be $15[\text{ms}]$. Therefore, for restraining the locus irregularity below $1[\mu m]$, it is necessary to set the reference input time interval of the digital controller below $\Delta T = 15[\text{ms}]$. For this purpose, it should prepare the computer which is capable of computing the velocity within $15[\text{ms}]$ for objective command calculation. In the industrial field, the allowance of locus irregularity is varied from the working aim. In the current NC machine tool, if the encoder resolution adopted in the motor is $1[\mu m]$ and its locus precision is required as $0.5[\mu m]$, the $10[\mu m]$ locus precision in laser cutting is needed. For guaranteeing this locus precision, the size of locus irregularity should be restrained to be a small value for satisfying the locus precision in the other reference input time intervals in the steady state. By using the relationship between reference input time interval and locus irregularity shown in Fig. 3.10, the reference input time interval ΔT can be determined based on the required locus precision in the design process. Fig. 3.10 can be also adopted as the useful figure in the design process.

4

Quantization Error of a Mechatronic Servo System

The control circuit of a servo controller is a completely software servo system equipped by software (micro-computer) and adopted widely in mechatronic systems in recent years. The rotation position of the motor is obtained from an encoder in the position detector installed in the motor. The resolution of the position is determined by a bit number of the encoder (encoder resolution). The quantization of torque information driving the motor (torque resolution) is determined by a D/A converter generating a command in the power amplifier according to motor current, equivalent to torque, and the bit number of the A/D conversion for performing feedback. In this chapter, encoder resolution and control performance of torque resolution and servo system is introduced.

4.1 Encoder Resolution

In the software servo system, a general velocity feedback signal is obtained according to the difference computation of the pulse signal about the position in encoder. When the encoder resolution is low, the resolution of velocity information then becomes low and contour control performance is degraded. In general, encoder resolution is determined from the positioning precision in many cases in industry. However, it is insufficient. Although it is necessary to determine the resolution considering contour control performance, the relation between the resolution of the encoder and control performance is not distinct in the past.

Concerning the software servo system, a mathematical model is derived while keeping the essential nature of encoder. Through analyzing this equation, the encoder resolution can be determined by equation (4.6) according to its contour control performance.

From contour control performance required in a software servo system, encoder resolution is determined properly.

4.1.1 Encoder Resolution of the Software Servo System

(1) Software Servo System Structure

In a software servo system, the position controller and the velocity controller are constructed in software. In addition, the current controller is also constructed in software. In this section, concerning the control problem about position, the position controller and velocity controller are only taken into account by neglecting the current controller and power amplifier whose properties are ideally considered. The relevant structure of the software servo system is shown in figure 4.1. The software servo system is briefly classified into the servo controller, motor and mechanism part. The position and velocity of motor are controlled by the servo controller. The control system of the servo controller is always constructed with the position loop and the velocity loop in the industrial field.

The positioning precision of a software servo system is determined by the resolution of the encoder installed in the servo motor, i.e., according to the measured position of the motor through dividing one rotation of the motor. The position output of the servo motor is the accumulated pulse output of the encoder by a counter, and measured by putting data at each sampling time interval (refer to section 3.1) into the servo controller.

In an analogue servo system, the velocity signal can be measured continuously by a velocity detector. In a software servo system, however, since the velocity detector is not installed, so as to reduce the cost, velocity is calculated from the position signal. Velocity calculation often adopted in the industrial field is the method according to the simple difference of position. In the following analysis, velocity computation is performed by difference computation. Since velocity can be only calculated based on resolution determined by the difference computation of the pulse in software servo system, the precision of velocity feedback is deteriorated compared with an analogue servo system. Hence, control performance is degraded due to a decrease of resolution of the velocity feedback signal because of difference computation, and ripple-type velocity fluctuation in the output of the the servo system is generated. This velocity fluctuation is different compared with the ripple-type velocity in the velocity detector of an analogue servo system. Since ripple-type velocity in

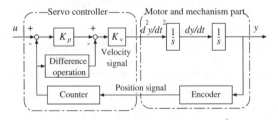

Fig. 4.1. Structure of software servo system

a velocity detector is generated as the detection noise of rotation velocity, ripple-type velocity can be prevented by smoothing this detection noise with a low pass filter. However, a ripple-type velocity fluctuation of a software servo system cannot be smoothed by a low pass filter because varied frequency introduced later is related with the objective velocity. Therefore, it is necessary to determine the encoder resolution for forcing the velocity fluctuation within the allowance region.

(2) Present Condition of Encoder Resolution Determination

The determination of present encoder resolution in the industrial mechatronic software serve system is carried out according to the necessity of positioning precision of a mechatronic servo system[4]. When performing contour control, the encoder resolution calculated from positioning precision is used without change. When required, control performance cannot be obtained, the encoder resolution with test error will be regulated. The determination of encoder resolution cannot be realized theoretically for the required control performance. Therefore, in this chapter, the theoretical determination method for encoder resolution for control performance, especially about contour control issue, considering the relationship between ripple-type velocity fluctuation and encoder resolution, is proposed.

4.1.2 A Mathematical Model and Resolution Judgement for Encoder Resolution

(1) A Mathematical Model of a Software Servo System

An industrial mechatronic servo system is always under the velocity condition of motion of the operated motor at $1/20 \sim 1/5$ of maximum velocity. Its dynamics is expressed by the 2nd order system as (refer to the 2.2.4)

$$Y(s) = \frac{K_p K_v}{s^2 + K_v s + K_p K_v} U(s) \tag{4.1}$$

where $Y(s)$ is the position output of the servo system, $U(s)$ is the position input of the servo system. K_p, K_v have the meaning of K_{p2}, K_{v2} in the equation (2.29) of the middle speed 2nd order model in the item 2.2.4, respectively.

The control system of the mechatronic servo system expressed by (4.1) is picked out from the software servo system shown in figure 4.1 for encoder resolution analysis. The model of software servo system for simplifying the analysis is shown in figure 4.2. From the structure of the software servo system (Fig. 4.1), the velocity feedback calculated based on difference computation is easily obtained from the external input. However, this external input, as a simple external input, is the same as the velocity signal in Fig. 4.1. This external input is the continuous feedback of the velocity output in an analogue

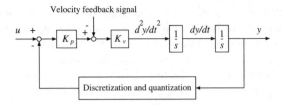

Fig. 4.2. Software servo system model for encoder resolution analysis

servo system. But in a software servo system, it is a discrete feedback. The basic unit of the position signal is 1[pulse]. The velocity signal is calculated with the difference computation of the position signal. The basic unit of the velocity signal according to difference computation is $1/\Delta t_p$[pulse/s], where Δt_p[s] is the sampling time.

(2) Relationship between Control Performance and Encoder Resolution

The relative equation between velocity fluctuation, occurred according to encoder resolution, and servo parameters is derived. In this part, the velocity fluctuation is analyzed when the motion of the servo motor is under the constant velocity, which is always adopted in the industrial field (refer to item 8 of 1.1.2). The flow of signal is as Fig. 1.1.2.

1. The difference divided according to velocity resolution $1/\Delta t_p$, determined by difference computation of the position signal, is accumulated. When the accumulated value is over the velocity resolution, the velocity feedback signal is added with $1/\Delta t_p$. This added velocity feedback signal is the reason for the velocity fluctuation.
2. According to the velocity loop gain K_v added into the velocity feedback signal, the input of the motor is varied with the step of $K_v/\Delta t_p$[pulse/s^2].
3. The change of velocity output of the motor based on the added velocity feedback signal is as $(K_v/\Delta t_p) \times \Delta t_p = K_v$[pulse/s], according to the integral of the input of the motor based on the sampling time interval Δt_p.

That is to say, the size of velocity fluctuation, occurred by the signal added into velocity feedback according to the effect of velocity resolution, is consistent with the value of velocity loop gain K_v. This relation can be expressed, if considering the unit, as

$$\Delta N = \frac{60 K_v}{R_E} \quad (4.2)$$

where ΔN[rev/min] denotes the velocity fluctuation amplitude with the ripple-type shape, R_E[pulse/rev] denotes the encoder resolution defined by the pulse number of the encoder when the motor rotates through one cycle.

This derived equation (4.2) is the fundamental equation for determining the following encoder resolution.

Next, the relationship between the velocity fluctuation period with the ripple-type shape and velocity of the objective trajectory is derived. If the velocity of the objective trajectory is as V_{ref}[pulse/s], the velocity feedback, obtained from the difference computation, is changed as $(\lceil V_{ref}\Delta t_p \rceil)/\Delta t_p$ when the velocity resolution is $1/\Delta t_p$, where $\lceil x \rceil$ is the maximal integer below x. From 1, this error is accumulated in each sampling time interval. Since the velocity fluctuation with the ripple-type shape occurred when the error is over $1/\Delta t_p$, The sampling time n at the moment of over $1/\Delta t_p$ is as

$$n\left(V_{ref} - \frac{\lceil V_{ref}\Delta t_p \rceil}{\Delta t_p}\right) = \frac{1}{\Delta t_p}. \qquad (4.3)$$

From (4.3), the velocity fluctuation frequency f_r[Hz] is calculated by

$$f_r = \frac{1}{n\Delta t_p} = \frac{V_{ref}\Delta t_p - \lceil V_{ref}\Delta t_p \rceil}{\Delta t_p}. \qquad (4.4)$$

From (4.4), the velocity fluctuation frequency f_r is depended on the velocity of objective trajectory V_{ref}. In order that the velocity fluctuation frequency f_r is not changed into a monotonic function about V_{ref}, a low pass filter cannot be adopted for smoothing.

(3) Determination of Encoder Resolution

By using (4.2), the relation equation between velocity fluctuation and encoder resolution derived by 4.1.2(2), the determination equation of the encoder resolution can be obtained. When the motor is rotated with a constant velocity, the ratio between the scale of the velocity fluctuation and the maximal velocity, called velocity fluctuation ratio R_N, is adopted as a specification of a mechatronic servo system, in order to express clearly the motion level of velocity of the motor. From this point of view, in the software servo system, the velocity fluctuation ratio R_N generated in the encoder resolution can be expressed by

$$R_N = \frac{\Delta N}{N_{max}} \qquad (4.5)$$

where, N_{max} denotes the maximal velocity [rev/min] of the servo motor. If we put (4.5) into (4.2), based on the solution of the encoder resolution R_E, the encoder resolution can be determined by

$$R_E = \frac{60 K_v}{R_N N_{max}}. \qquad (4.6)$$

The equation (4.6) is the final derived result in this section. According to this equation, proper encoder resolution R_E can be decided for satisfying the velocity fluctuation ratio R_N, determined according to the application of the servo motor from the maximal velocity N_{max} and velocity loop gain K_v.

84   4 Quantization Error of a Mechatronic Servo System

4.1.3 Experimental Verification of the Encoder Resolution Determination

(1) Experimental Verification of the Relationship between the Encoder Resolution and Control Performance

From the experiment, the relationship between the encoder resolution and the velocity fluctuation is verified. In the experiment, DEC-1(refer to the experiment deviceE.1) was adopted. Actually, DEC-1 was originally constructed with an analogue servo system. However, in this experiment, a software servo system using a computer was used. That is to say, the pulse output of the servo motor is accumulated by a counter equipped in the computer. The computer program implements the servo controller. Its output is put into servo

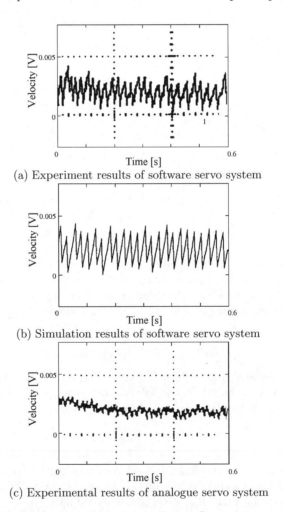

Fig. 4.3. Verification of velocity ripple in software servo system

amplifier by using a D/A converter for constructing the software servo system. The resolution of D/A conversion is adopted with reduction by a 1/100 amplifier from the D/A converter, which can permit ±5[V] with a resolution of 12[bit]. Since 1[bit] is about 2.44×10^{-5}[V], the effect of resolution to control performance can be neglected. In addition, the velocity of the servo motor is measured using digital data storage providing velocity detector (tachogenerator) output equipped with a load generator. This tachogenerator output is 7[V] with a rotational frequency of 1000[rev/min] of the servo motor. Since there are many factors of noises in the tachogenerator, the 100[Hz] low pass filter is adopted to eliminate these noise factors. The resolution of the encoder installed in the servo motor is 2000[pulse/rev]. But from the tested two increase and decrease signals of the encoder output as putting them into the pulse counter, the original 1[pulse] is changed into 4[pulse]. Through the 4 times circuit, it can be obtained as $R_E = 8000$[pulse/rev]. The maximal velocity is $N_{\max} = 1000$[rev/min], the sampling time interval $\Delta t_p = 4$[ms] (refer to 3.1). The position loop gain and velocity loop gain are set as $K_p = 12$[1/s] and $K_v = 68$[1/s] so that there is no oscillation or overshoot in the analogue servo system (refer to 2.1.2). Since velocity fluctuation is one of the problems in the industrial field, for big velocity fluctuation in low speed, ramp input for DEC 1 is $u(t) = 40t$[pulse], i.e., rotation speed of motor is 0.3[rev/min] for low speed. In the steady state, the experimental results and simulation results are illustrated in Fig. 4.3. From Fig.(a), (b), in the steady state, the amplitude in experimental results and in simulation results are both 0.004[V]. The frequency in both about is 40[Hz]. The shape of the waves are both triangular. From the above, it can be verified that the experimental results and simulation results are almost the same. In Fig.(a) of experimental results, the size of velocity fluctuation is about 0.004[V], i.e., 0.57[rev/min]. This value is almost the same as the size $\Delta N = 60 \times 68/8000 = 0.51$[rev/min] of velocity fluctuation calculated by equation (4.2). In addition, the velocity fluctuation frequency is also consistent with the frequency 40[Hz] calculated by equation (4.4).

To verify, the experimental results of an analogue servo system with same conditions are illustrated in Fig.(c). In the analogue servo system, the velocity fluctuation does not occur at all. The velocity fluctuation in Fig.(a) is verified that it is the cause of the resolution of software servo system by the experiment of 4.1.2(2).

(2) Application of Encoder Resolution Determination

Using equation (4.6) derived by 4.1.2(3), the example of determining the encoder resolution is illustrated. In DEC-1 adopted in the previous experiment, the necessary encoder resolution is $R_E = 60 \times 68 \times 1000/1000 = 4080$[pulse/rev] obtained from equation (4.6) if the velocity fluctuation ration is given as $R_N = 1 \times 10^{-3}$. In contrast, if the installed encoder resolution is actually $R_E = 8000$[pulse/rev], the velocity fluctuation ratio is

86   4 Quantization Error of a Mechatronic Servo System

$R_N = 60 \times 68/(1000 \times 8000) = 5.1 \times 10^{-4}$. From this point of view, according to the encoder resolution determination equation (4.6), the encoder resolution can be easily determined from the required velocity fluctuation ratio.

4.2 Torque Resolution

In the software servo system, the feedback of the motor current equivalent to the torque is carried out through a micro-computer. Between the power amplifier for driving the motor and the micro-computer is the A/D, D/A conversion. The theoretical relation between the A/D, D/A conversion quantization error and control performance must be clarified.

The appropriate mathematical model for the relationship between the torque resolution of the software servo system and control performance is derived. According to the solution of the mathematical model, the positioning precision by equation (4.8) and the position fluctuation of the ramp response by equation (4.15)~(4.17), with regard to the torque resolution, can be clarified. According to the bit number proposed in the A/D, D/A converter, the control performance of the servo system can be clearly estimated. Additionally, the minimal necessary bit number of the D/A, A/D conversion for testing out torque command and current feedback, in order to implement the necessary control performance of the software servo system, can be determined by equation (4.25).

4.2.1 Mathematical Model of the Mechatronic Servo System for Torque Resolution

The conceptual graph of the discussed software servo system in this section is shown in Fig. 4.4. The software servo system is shown in Fig. 4.4. In order to construct the control circuit of the servo controller using micro-computer software, the torque (current) command output from the control circuit is quantized. Therefore, the current reference input to the power amplifier actually needs a D/A converter. The block diagram of the 2nd order system of the servo system including torque quantization is illustrated by Fig. 4.5. $K_p[1/s]$, $K_v[1/s]$ have the meanings of K_{p2}, K_{v2} in the middle speed 2nd order model equation (2.29) of item 2.2.4. In addition, the sampling time interval of the velocity loop is $\Delta t_v[s]$. The servo system is usually constructed with position feedback, velocity feedback and current feedback. The position feedback and velocity feedback refer to the feedback of the actual motor output for the servo controller. The current feedback refers to the feedback of power amplified. It is not changed into the actual torque. For the mathematical model of the servo system in the block diagram of Fig. 4.5, the position feedback and velocity feedback is widely considered. The current feedback is simply assumed as the output of the power amplifier. The control method of the velocity loop is P control or PI control. But the entire property of the velocity

4.2 Torque Resolution   87

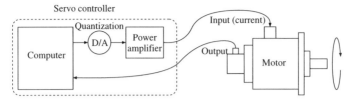

Fig. 4.4. Structure of software servo system

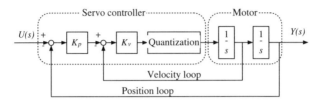

Fig. 4.5. The 2nd order model of software servo system including torque quantization

loop is expressed by the 1st order system. The position control and velocity control are combined into the 2nd order system (refer to item 2.2.4).

In this section, the torque quantization with A/D, D/A conversion, as a problem, is expressed according to the quantization term in Fig. 4.5. By the function $f(\cdot)$ for quantization of torque, the mathematical model of a servo system including the torque quantization is as

$$\frac{d^2y(t)}{dt^2} = f\left(K_pK_vu(t) - K_pK_vy(t) - K_v\frac{dy(t)}{dt}\right). \tag{4.7}$$

For measuring the rotation angle of the servo motor by a pulse [pulse] according to the encoder, the rotation angle u of motor as a position command is expressed by a pulse. The angular velocity input, as the velocity command, is $K_p\{u(t) - y(t)\}$[pulse/s]. The angular acceleration input, regarded as the torque command to torque quantization, is $K_v[K_p\{u(t) - y(t)\} - dy(t)/dt]$[pulse/s^2]. In order to make the angular acceleration quantization function $f(x)$ as the step-wise function of Fig. 4.6, the input angular acceleration x[pulse/s^2] is quantized by the angular acceleration resolution R_A[pulse/s^2].

In addition, considering the effect of torque quantization on the control performance, it assumed that position and velocity without quantization are feedback with continuous values. In the actual software servo system, the encoder resolution of the servo motor is infinite. That is, the position and velocity information is continuously obtained at the desired state. Compared with the actual software servo system with an encoder, the control performance with this assumption is the maximum possible. The condition of deriving torque resolution is considered as the prerequisite condition. In the software servo sys-

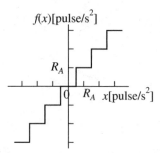

Fig. 4.6. Quantization of angular acceleration

tem, for realizing the required control performance, the A/D, D/A conversion is carried out with torque resolution capable of satisfying the lower limitation. Moreover, since introducing this assumption, the analysis of the effect on the control performance of torque resolution becomes easy and it is possible to derive the torque resolution condition equation by 4.2.4(1), (2). The appropriation of this condition equation in 4.2.4(4) is completely expressed by a computer simulation taking into account the encoder of the servo motor.

4.2.2 Deterioration of Positioning Precision Due to Torque Quantization Error

(1) Position Determination of the Software Servo System

For determining the position of the software servo system, the effect of the torque quantization error is considered. The positioning error $E_p^s = P_{ref} - y(\infty)$[pulse], which is the error of objective position P_{ref}[pulse] and the steady-state value of the position output $y(\infty)$[pulse], is determined based on the servo parameter K_p, K_v and the angular acceleration resolution. The relationship equation is derived theoretically. As illustrated in Fig. 4.7, the servo motor is rotated with a constant velocity input according to the objective position P_{ref}. The position can be determined. If the angular acceleration R_A is quantized, the velocity of the servo motor will be also quantized in each sampling time interval Δt_v of the velocity loop. That is, in the servo system with the angular acceleration quantization, the velocity output is only changed with the unit of $R_A \Delta t_v$[pulse/s]. This quantized resolution is called the angular velocity resolution. From this case, for the servo system with angular acceleration quantization, the velocity feedback is carried out until that angular velocity output becomes 0[pulse/s]. When the angular velocity output becomes zero, the velocity feedback is cut off and the steady state is continued until the position output becomes constant.

(2) Relationship between Positioning Error and Angular Acceleration Resolution

At the moment that the input is equal to the objective position P_{ref}, the input to the quantization term of Fig. 4.5 is expressed by $K_v(K_p(P_{ref} - y) - dy/dt)$. When this value larger than the angular acceleration resolution R_A, the position and velocity is feedback. If the angular acceleration resolution is not full, that is, $dy/dt = 0$[pulse/s], the output of the quantization term is 0 and the position output remains constant.

In the steady state that the position output is constant, the size of the input to the quantization term is expressed by $|K_p K_v E_p^s|$ with the positioning error E_p^s, as Fig. 4.7. When this value is not full of resolution R_A of the angular acceleration, the position error E_p^s can be expressed by K_p, K_v, R_A as

$$|E_p^s| < \frac{R_A}{K_p K_v}. \tag{4.8}$$

From (4.8), the upper limit of the position error E_p^s is proportional with the angular acceleration resolution R_A and inversely proportional to the position, velocity loop gain K_p, K_v.

4.2.3 Deterioration of Ramp Response Due to Torque Quantization Error

(1) Ramp Response of the Software Servo System

Next, with regard to the ramp input of the software servo system, the effect of torque quantization error is considered. The objective trajectory of the servo motor is given with the constant velocity V_{ref}[pulse/s]. When the angular acceleration is quantized in each R_A, if the objective angular velocity is the integer times of the angular velocity resolution, the angular velocity output is not changed for making the objective angular velocity consistent with angular velocity output. However, if the objective angular velocity is not the

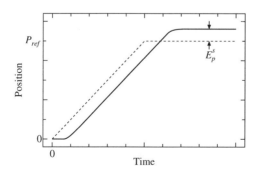

Fig. 4.7. Deterioration of position control in software servo system

90   4 Quantization Error of a Mechatronic Servo System

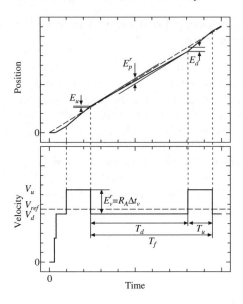

Fig. 4.8. Deterioration of ramp response in software servo system

integer times of the angular velocity resolution, the angular velocity output is changed because of inconsistence between objective angular velocity and angular velocity output.

Fig. 4.8 illustrated the variation of the angular velocity output. The upper part of Fig. 4.8 shows the position fluctuation and the bottom part shows the angular velocity fluctuation. From Fig. 4.8, the response is divided into two states: one is that the angular velocity output is below the objective angular velocity (scale of T_d[s]) and another is that the angular velocity output is over the objective angular velocity (scale of T_u[s]).

(2) State of Angular Velocity Output under Objective Angular Velocity V_{ref}

At the state of that the angular velocity output is below the objective angular velocity V_{ref}, from the angular velocity quantization, the output angular velocity is as $V_d = \lceil V_{ref}/(R_A \Delta t_v) \rceil R_A \Delta t_v$[pulse/s] (where $\lceil x \rceil$ is expressed as the maximal integer below x). The error $V_{ref} - V_d$ between objective angular velocity and angular velocity output is made integral as the position output error. If the angular acceleration input is over half of the angular acceleration resolution $R_A/2$ (refer to Fig. 4.6), the positive pulse equivalent to the angular acceleration resolution is generated. When generating the pulse and the position output error is E_d[pulse], the angular acceleration input is expressed as $K_v(K_p E_d - V_d)$ with the loop of Fig. 4.5. When this value is half of the angular acceleration resolution $R_A/2$, the following relationship equation is successful

$$K_v(K_p E_d - V_d) = \frac{R_A}{2}. \tag{4.9}$$

If solving the equation (4.9) for E_d, it is as

$$E_d = \frac{R_A + 2K_v V_d}{2K_p K_v}. \tag{4.10}$$

The amplitude of the position error is in the positive direction in equation (4.10). Additionally, the amplitude of this velocity error is as $V_{ref} - V_d$.

(3) State of Angular Velocity Output over Objective Angular Velocity V_{ref}

If generating the pulse equivalent to the angular acceleration resolution, the angular velocity output $R_A \Delta t_v$ is increased as $V_u = \lceil V_{ref}/(R_A \Delta t_v) + 1 \rceil R_A \Delta t_v$ [pulse/s]. The error $V_u - V_{ref}$ of the objective angular velocity and angular velocity output is made integral as the position output error. If the angular acceleration input is over half of the angular acceleration resolution $R_A/2$ (refer to Fig. 4.6), the negative pulse equivalent to the angular acceleration resolution is generated. If generating the pulse and the position output error is E_u [pulse], the angular acceleration input is expressed as $-K_v(K_p E_u + V_u)$. If this value is half of the angular acceleration resolution $R_A/2$, the following relation equation is successful

$$-K_v(K_p E_u + V_u) = -\frac{R_A}{2}. \tag{4.11}$$

If solving the equation (4.11) for E_u, it is as

$$E_u = \frac{R_A - 2K_v V_u}{2K_p K_v}. \tag{4.12}$$

The amplitude of the position error is in the positive direction in equation (4.12). Moreover, the amplitude of the velocity error is as $V_u - V_{ref}$. When generating this negative pulse, the angular velocity output is back to the state that the angular velocity output of 4.2.3(2) is over the objective angular velocity. Because of these two states, a fluctuation of the ramp response exists.

(4) Amplitude and Cycle of Position Fluctuation

From Fig. 4.8, if the position has deviation $E_u + E_d$, when the error between the objective velocity and tracing velocity $V_{ref} - V_d$ is continued at the time T_d, the time T_d of that the angular velocity output is continuously below the objective angular velocity is as

$$T_d = \frac{E_d + E_u}{V_{ref} - V_d}$$
$$= \frac{R_A(1 - K_v \Delta t_v)}{K_p K_v (V_{ref} - V_d)} \tag{4.13}$$

by using equation (4.10) and (4.12). Similarly, the time T_u of that is the angular velocity output is continuously over the objective angular velocity, is as

$$T_u = \frac{E_d + E_u}{V_u - V_{ref}}$$
$$= \frac{R_A(1 - K_v \Delta t_v)}{K_p K_v (V_u - V_{ref})}. \tag{4.14}$$

The fluctuation period $T_f[s]$ is as

$$T_f = T_d + T_u$$
$$= \frac{R_A^2 \Delta t_v (1 - K_v \Delta t_v)}{K_p K_v (V_{ref} - V_d)(V_u - V_{ref})} \tag{4.15}$$

by combining the T_d of equation (4.13) and the T_u of equation (4.14). The amplitude of position fluctuation E_p^r[pulse] is as

$$E_p^r = E_d + E_u$$
$$= \frac{R_A + K_v(V_d - V_u)}{K_p K_v}$$
$$= \frac{R_A(1 - K_v \Delta t_v)}{K_p K_v} \tag{4.16}$$

by combining the equation (4.10) and equation (4.12). The amplitude of the velocity E_v^r[pulse/s] is as

$$E_v^r = R_A \Delta t_v \tag{4.17}$$

from the angular acceleration resolution.

From above derived equation (4.15), the equation (4.17) is the relationship equation expressing the relation among fluctuation period T_f, amplitude of position fluctuation E_p^r, velocity fluctuation amplitude E_v^r and angular acceleration resolution R_A. By analyzing their properties from these relation equations, the amplitude of position fluctuation E_p^r and the velocity fluctuation amplitude E_v^r are proportional to the angular acceleration resolution R_A, and the fluctuation period T_f proportional to twice that of the angular acceleration resolution R_A. In addition, the fluctuation period T_f is dependent on the objective velocity V_{ref}. The amplitude of position fluctuation E_p^r and the velocity fluctuation amplitude E_v^r are not reliant on the objective velocity V_{ref}. The angular acceleration resolution R_A is dependent on the parameter K_p, K_v, Δt_v of the servo controller.

4.2.4 Derivation of Torque Resolution Determination

(1) Positioning Precision

If the positioning error output E_p^s is not full of 1[pulse] of encoder output, there is no effect of torque resolution compared with the encoder resolution. If the right side of equation (4.8) is not full of 1 as

$$|E_p^s| < \frac{R_A}{K_p K_v} < 1 \tag{4.18}$$

and by solving the R_A, the condition of angular acceleration resolution R_A if the position error E_p^s is not full of 1 can be expressed as

$$R_A < K_p K_v. \tag{4.19}$$

That is, in order to make the positioning precision E_p^s of the servo motor is full of 1, the angular acceleration resolution R_A should be determined by satisfying the equation (4.19).

(2) Fluctuation of Ramp Response

In the ramp response, the torque resolution is determined when the amplitude of angular velocity output deterioration E_v^r and the amplitude of position output deterioration E_p^r are within the allowance E_{limv}^r[pulse/s] and E_{limp}^r[pulse], respectively. The upper bound R_{Ap}[pulse/s^2] of the angular acceleration resolution satisfying the condition of amplitude of position output deterioration can be calculated using equation (4.16)

$$R_{Ap} = \frac{K_p K_v E_{limp}^r}{1 - K_v \Delta t_v}. \tag{4.20}$$

The upper bound R_{Av}[pulse/s^2] of the angular acceleration resolution satisfying the condition of the amplitude of the angular velocity output deterioration can be calculated using equation (4.17)

$$R_{Av} = \frac{E_{limv}^r}{\Delta t_v}. \tag{4.21}$$

The angular acceleration resolution R_A is needed from equation (4.20) and equation (4.21)

$$R_A \leq \min(R_{Av}, R_{Ap}). \tag{4.22}$$

That is, when the angular acceleration resolution R_A can be determined for satisfying the equation (4.22), the restraint of the deterioration of ramp response within the demanded allowance can be realized.

(3) Calculation of Bit Numbers of torque resolution

The angular acceleration resolution R_A should correspond to the bit number of the D/A, A/D conversion which is adopted for the current reference and feedback of the software servo system. First of all, the angular acceleration resolution is converted into the torque resolution R_T[Nm] using the pulse number P[pulse/rev] equivalent to the moment of inertia J_M[kgm^2] of the motor and the encoder of one time rotation.

$$R_T = \frac{2\pi R_A J_M}{P}. \tag{4.23}$$

Next, the bit number of the A/D, D/A conversion should be converted. That is, the maximum of the bit number of the A/D, D/A conversion adopted for the software servo system must be able to output the maximum torque of the motor except the symbol bit. The relation equation amongst the bit number B[bit] of the resolution of the A/D, D/A conversion, the maximum torque $T_{\max}$[Nm] of the maximum torque of motor and the torque resolution R_T[Nm] should be given as

$$\begin{aligned} 2^{B-1} &= \frac{T_{\max}}{R_T} \\ &= \frac{T_{\max} P}{2\pi R_A J_M}. \end{aligned} \tag{4.24}$$

The final conversion is based on equation (4.23). The equation (4.24) is changed after solution about bit number B of the resolution of the A/D, D/A conversion as

$$B = \log_2 \frac{T_{\max} P}{\pi R_A J_M}. \tag{4.25}$$

By using the bit number from equation (4.25), the performance of a A/D, D/A conversion satisfying the demanded precision of the ramp response can be determined.

(4) Numerical Example of Torque Resolution Determination

The effectiveness of using the relationship between the derived control performance of the software servo system and the bit number of the A/D, D/A conversion in the software servo system is verified here.

The designed position loop gain and the velocity loop gain of the servo controller are $K_p = 40[1/s]$ and $K_v = 200[1/s]$, respectively. The sampling time interval for the velocity loop is $\Delta t_v = 50[\mu s]$. The rated values of the servo motor are $J_M = 0.13 \times 10^{-4}$[kgm^2], $T_{\max} = 1.47$[Nm], $P = 5000$[pulse/rev]. For making the positioning precision as (4.19), the deterioration of the ramp response as $E^r_{limp} = 1$[pulse] and $E^r_{limv} = 1$[pulse/s], the angular acceleration resolution is determined by equation (4.22), and the calculation of the bit number B of the torque resolution using equation (4.25) for $B = 15$[bit].

When existing quantization of the position information in the actual software servo system, in order to investigate the degree of the obtained the control performance based on the torque resolution, the computer simulation is made using torque resolution and considering the quantization of position information. The objective trajectory is $u(t) = 10000t$[pulse] $(0 \leq t < 1\text{[s]})$, $u(t) = 10000$[pulse] $(1 \leq t \leq 2\text{[s]})$. The approximation of the velocity information is based on the discrete error of the position information. Additionally, the positioning precision is $E_p^s = 1$[pulse], the position fluctuation of the ramp response is $E_p^r = 2$[pulse] and the velocity fluctuation is $E_v^r = 200$[pulse/s]. Even without considering the torque resolution and only considering the quantization of the position information, all of the positioning precision, the position fluctuation of the ramp response and the velocity fluctuation have the same values. When existing quantization of the position, the effect of torque quantization in the derived torque resolution can be neglected for making the results consistent between considering the torque quantization and not considering the torque quantization. Moreover, in the desired state without considering the quantization of position, if comparing the design values and simulation results, the expected control performance can be obtained by using the derived torque resolution about the positioning precision and the position fluctuation of the ramp response. The reason for deterioration of the design values caused by velocity fluctuation is that the position information is simply discrete when approximating the velocity information. That is, the resolution of the velocity approximation values is as $1[\text{pulse}]/\Delta t_v[\text{s}] = 1/(50 \times 10^{-6}) = 20000[\text{pulse/s}]$ when approximating the velocity information based on the discrete values. If standardizing the resolution of the velocity approximation value, the velocity fluctuation is very small at 1%.

Next, in order to calculate the bit number of the torque resolution satisfying the general required control performance, the relationship equation among the positioning error E_p^s expressed by equation (4.8), the amplitude of position fluctuation E_p^r expressed by equation (4.16), the velocity fluctuation amplitude E_v^r expressed by equation (4.17), and the angular acceleration resolution R_A can be converted into the bit number B of the torque resolution by equation (4.25). This relationship is shown in Fig. 4.9.

According to the use of Fig. 4.9, although the bit number of the torque resolution cannot be worked out from the required control performance, the positioning precision and the control performance of the ramp response can be obtained from the bit number of the torque resolution of the actual operated software servo system.

(5) Relationship Among Control Performance, Torque Resolution and Servo Parameter

The relationship amongst the control performance, torque resolution and servo parameter of the software servo system is summarized as below.

96   4 Quantization Error of a Mechatronic Servo System

1. As shown in Fig. 4.9, the logarithm of positioning error E_p^s, the amplitude of position fluctuation E_p^r, the velocity fluctuation amplitude E_v^r and the bit number B of the torque resolution are expressed by a 1st order function.
2. From equation (4.8) and equation (4.16), the positioning error E_p^s and the amplitude of position fluctuation E_p^r of the ramp response have the negative proportion with the position loop gain K_p.
3. From equation (4.17), the velocity fluctuation amplitude E_v^r is not dependent on the servo parameter K_p, K_v.
4. From equation (4.15), the fluctuation period T_f of ramp response is dependent on the objective velocity V_{ref}.

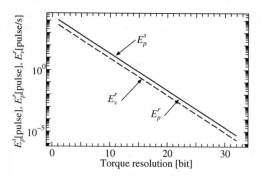

Fig. 4.9. Relation between the resolution of A/D, D/A conversion and control performance

5

Torque Saturation of a Mechatronic Servo System

In order to realize the high-speed motion of a mechatronic system, the power amplifier current for driving the torque of servo motor is adopted. This often generates a deterioration of contour control performance when occurring torque saturation. In this chapter, the measurement method of torque saturation and a simple algorithm of generating the motion velocity for preventing the deterioration of contour control performance, when the torque saturation properties are known, is shown.

5.1 Measurement Method for the Torque Saturation Property

The previous torque saturation property refers to the use of a power amplifier property without any change. The saturation property of a power amplifier is defined by measuring the relationship of the output signal with the simple amplifier input signal. However, in a mechatronic servo system, the power amplifier is installed in the servo controller and it drives the motor connecting with the mechanism part. When analyzing the contour control performance of this mechatronic servo system, the torque saturation property of the mechatronic servo system cannot be reflected sufficiently when only considering the saturation property of the power amplifier and it is also difficult to grasp the performance deterioration quantitatively.

It will introduce the calculation method of the torque saturation property according to the comparison between the output of a linear mathematical model expressing the motion property of a mechatronic servo system and the actual output.

Since the torque saturation property can be calculated through the proper expressed mathematical model on the dynamics of a mechatronic servo system during the contour control, the torque saturation property can be adopted effectively to improve the contour control performance of a mechatronic servo system.

5.1.1 Torque Saturation of a Mechatronic Servo System

(1) Torque Saturation Problem in the Mechatronic Servo System and Necessity of Torque Saturation Property Measurement

Torque saturation, as a big problem, is the reason for performance deterioration when performing high-speed contour control of a mechatronic system. Torque saturation refers to the phenomenon generated according to the properties of a servo controller and motor and also the phenomenon that the response torque cannot be generated actually corresponding to the given big torque command in the mechatronic servo system. Torque saturation can occur under the movement with 1/5 of the rated speed. It is not a special phenomenon that will be generated near the highest velocity. In the contour control of an industrial mechatronic servo system, if torque saturation occurs, there will appear to be an obvious deviation between objective trajectory and the following trajectory. Therefore, torque saturation is absolutely not permitted to occur.

By summarizing the items related to the torque of an industrial mechatronic servo system, there are following items.

1. Servo stops with overload current: if the current is over the maximum allowable current, which is equal to the torque of the transient motor output (3 $\sim$ 5 times of rated torque), flows into the motor over 1[s], the servo will stop with a current interruption. Namely, the servo will stop when the current is more than the equivalent rated torque flows for a certain time. They are the measures to avoid motor damage from overload current.
2. Wind up measure: when there are many PI control systems used in the current control part and the velocity control part, a wind-up phenomenon will occur at the saturation of the integral (I) action. The wind-up phenomenon is the phenomenon that the output cannot be decreased along with the input decrease because the controller continuously increases the deviation of a certain symbol according to the difference between the required manipulation and actual manipulation after saturation, when manipulation contains saturation in the controller including the integrator. PI control part of the current control part is designed so that it absolutely does not enter into the saturation region in the whole control period, and several strategies, such as current interruption, etc., are adopted to stop the servo when the PI control part passing through the fixed period of entering into the saturation region. Concerning the PI controller of the velocity control part, it will perform when the current reference, as an output of the PI controller, is within the torque which can be output by the motor. However, when the current reference is over the torque which can be output by the motor, integral (I) action will be divided and P control will be performed to avoid the wind-up phenomenon.

5.1 Measurement Method for the Torque Saturation Property   99

3. Counter-electromotive force compensation: the current value, which can be adopted in the torque of acceleration-deceleration from the difference between the terminal voltage input added into the motor and counter-electromotive force of the motor, can be guaranteed by counter-electromotive force compensation. This counter-electromotive force compensation is a method that the current of the motor will be flown with $3 \sim 5$ times of the rated torque even in high-speed rotational frequency, when the current is equivalent to a large terminal input voltage of the motor in the PWM amplifier corresponding to the counter-electromotive force in the high-speed rotational frequency of the motor.

In the industrial mechatronic servo system with the above features, torque saturation will occur in the mechatronic system due to the saturation feature of the current output of the power amplifier or torque output of the servo motor. Around this saturation, the movement of the mechatronic servo system is very difficult. Taking into account the safety of the equipment, the clip (restriction) should be equipped in the velocity control part and current control part in the servo controller before saturation occurs. The clip of the velocity control part is the variation of the velocity, i.e., variation of velocity not over acceleration. The clip of the current control part is the value which is not over the current value. Both of above clips exist not over the current variation. From these clips, restriction is extorted for the torque of the mechatronic servo system. In this chapter, torque output with restriction according to the clip is called torque saturation.

In the current situation, when high-precision contour control of the mechatronic servo system is required, it can be realized in the linear region of low velocity without generation of torque saturation. Nevertheless, the high-speed motion is always demanded. Therefore, it is very important to grasp the torque saturation property in the high-precision contour control. If understanding the torque saturation property correctly, it is possible to realize high-precision contour control performance for objective velocity in the possible limited velocity according to the torque saturation property.

In the following analysis, discussion will be carried out for the servo system with one axis. The industrial mechatronic system is composed of each independently controlled axis. Also, in the articulated robot arm using nonlinear kinematics between working coordinates and joint coordinates, for making the servo system of each axis with basic structure exactly the same, the general torque saturation of an industrial mechatronic system can be regarded as the issue of one axis (refer to 1.1.2 item 6).

(2) Analysis Method of Reasons for Load Torque Generation in the Mechatronic Servo System and Derivation Method of the Torque Saturation Property

In the industrial mechatronic servo system, the biggest load torque is coulomb friction and next is viscous friction. The generation cause of coulomb friction and viscous friction lies in the frictional part or the sleeve part. The intensity of friction is variable from the tightening of bearing or adopted grease. Comparing with the friction of the motor itself, friction of the mechanism is dominant. In the industrial mechatronic servo system, these frictions often compose 60% ∼70% of the rated torque. Coulomb friction is determined by the constructed status of the mechanism part. Viscous friction can be considered as a certain load torque for not causing large-scale velocity variation in the movement status.

In the case of an articulated robot arm, generally, there is no constant velocity in joint coordinates even with a constant velocity in working coordinates, because of nonlinear coordinate transforms between working coordinates and joint coordinates. However, in the working field generally adopted in contour control and working status, constant velocity should be set in order to have no great change of velocity of the motor. Besides, the motor axis equivalent moment of inertial should be changed along with the change of attitude of the robot and so on. But these torque variations introduced above is smaller than the frictional torque in the normal scale of movement. In addition, most of the load of the servo motor is loaded from the mechanism part. The mechanism part cannot be executed with non-load operation, but often can be moved around the rated load in the status of a mechatronic servo system after construction. Thus, discussion with the condition of a certain load torque for the entirety can be carried out.

From the view of the contour control of a mechatronic servo system, the torque saturation property is not a saturation property of the power amplifier itself. Concerning the impact of the torque saturation property on contour control, the expected torque output with an assumption of no torque saturation is defined as standard. If the relation with the generated actual torque can be measured according to the impact of torque saturation of the system, the torque saturation property can be obtained from contour control performance. Namely, according to the calculated torque output based on a linear model expressing correctly dynamics of a mechatronic servo system when performing contour control without torque saturation, the torque saturation curve can be worked out by comparing it with the torque output of an actual mechatronic servo system. In the following part, a new measurement method of the torque saturation property will be described from the above contour control point of view.

(3) Measurement Method of the Torque Saturation Property Based on the Comparison with the Desired Conditions

Based on the analysis method of 5.1.1(2), the derivation method of the torque saturation property of an actual mechatronic servo system is introduced. First, the velocity step response of the actual mechatronic servo system is worked out respectively. This step response of velocity is consistent with the velocity response in the expected status without torque saturation at low speeds.

5.1 Measurement Method for the Torque Saturation Property

However, due to the impact of torque saturation when the step response of velocity is over a limited value, the response of an actual system is disappeared as the expectation. In order to understand these characteristics, the procedure of the measurement method of the torque saturation property, based on the comparison with the theoretical value, is shown. In the following explanation on the procedure of measurement method, the discussion is carried out using the acceleration-deceleration in the proportion relation with torque adopted in acceleration-deceleration. The procedure of measurement for the torque saturation is summarized as below.

1. Velocity output is measured at a given velocity step input of actual system.
2. Acceleration output is calculated based on the information of the velocity output.
3. Maximal acceleration output is calculated from the acceleration output wave.
4. To a given velocity command, the theoretical acceleration output is worked out based on the linear model expressing the dynamics of system with an assumption of no torque saturation.
5. A graph is plotted with horizontal axis showing the theoretical value of the acceleration output at the status calculated in 4, and the vertical axis representing the acceleration output measured from the actual system calculated in 3.
6. Expanding to the whole velocity region used in contour control, the typical points of acceleration saturation properties is worked out by repeating steps 1∼5 for different velocity step input values. By interpolating between these points, the torque saturation curve is drawn out.

According to the above steps, the acceleration saturation property can be understood. For the proportional relationship between acceleration and torque adopted in acceleration-deceleration, this graph can express the torque saturation when the motor axis equivalent moment of inertia in the acceleration is drawn by curve and the torque which cannot occur linearly according to the expectation is graphed. The torque saturation property worked out by this way, i.e., the upper bound and lower bound of torque saturation cannot be found and torque is always linearly output even in the region, in which saturation does not occur.

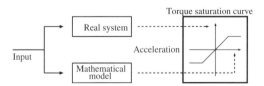

Fig. 5.1. Conceptual graph of torque saturation property derivation

102   5 Torque Saturation of a Mechatronic Servo System

By using a mathematical model expressing the dynamics of the system when torque saturation does not occur as a desired status, the theoretical acceleration output needed in step 1 of measure procedure, is explained.

The conceptual graph of torque saturation property 5.1 is shown concretely in Fig. 5.2. The adopted mathematical model is the linear 2nd order model expressing the property in the state of without torque saturation in a mechatronic servo system. It is illustrated in the bottom part of Fig. 5.2.

To position input $U(s)$ of the mechatronic servo system, dynamics of acceleration output $s^2Y(s)$ is expressed as (refer to item 2.2.4).

$$s^2Y(s) = \frac{K_p K_v s^2}{s^2 + K_v s + K_p K_v} U(s) \quad (5.1)$$

where K_p, K_v have the meanings as K_{p2}, K_{v2} in the middle speed 2nd order model as equation (2.29) in item 2.2.4, respectively. $s^2Y(s)$ describes the Laplace transform of the acceleration output (refer to appendix A.1). Since the command in a mechatronic servo system is given with a designated velocity in contour control, it is necessary to measure the saturation property under the same conditions for working out torque saturation. Acceleration output is calculated if input $U(s)$ is a fixed velocity as

$$U(s) = \frac{v}{s^2}. \quad (5.2)$$

Acceleration can be worked out according to the following simple analytical solution. The mechatronic servo system in the industrial field generally has two different real poles in equation (5.1) according to the relationship between position loop gain K_p and velocity loop gain K_v (refer to item 2.2.4). To

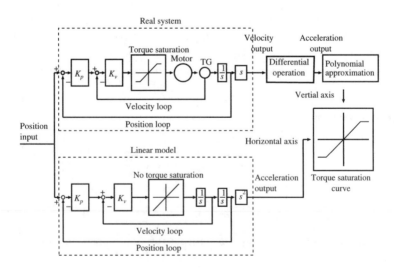

Fig. 5.2. Block diagram of corner acceleration saturation property measurement

5.1 Measurement Method for the Torque Saturation Property

velocity step input of equation (5.2) of a mechatronic servo system, equation (5.1) is input into equation (5.2). Through a inverse Laplace transform (refer to appendix A.1), the acceleration output can be calculated by

$$\frac{d^2 y(t)}{dt^2} = \frac{K_p K_v v}{p_1^s - p_2^s} e^{p_1^s t} - \frac{K_p K_v v}{p_1^s - p_2^s} e^{p_2^s t} \qquad (5.3)$$

$$p_1^s = -\frac{K_v + \sqrt{K_v^2 - 4 K_v K_p}}{2}$$

$$p_2^s = -\frac{K_v - \sqrt{K_v^2 - 4 K_v K_p}}{2}.$$

At this time, in order to calculate the maximum value of torque, the differential of equation (5.3) is as

$$\frac{d^3 y(t)}{dt^3} = -\frac{K_p K_v v p_1^s}{p_2^s - p_1^s} e^{p_1^s t} + \frac{K_p K_v v p_2^s}{p_2^s - p_1^s} e^{p_2^s t}. \qquad (5.4)$$

The moment t_M, when the acceleration of equation (5.3) reaches a maximum value, can be obtained from the solution of $d^3 y(t)/dt^3 = 0$ as

$$t_M = \frac{1}{p_2^s - p_1^s} \log \frac{p_1^s}{p_2^s}. \qquad (5.5)$$

Furthermore, for the step response of the velocity, i.e. about acceleration, on velocity, acceleration output peak of impulse response (maximum acceleration) can be worked out by

$$\frac{d^2 y(t_M)}{dt^2} = \frac{K_p K_v v}{p_1^s - p_2^s} e^{p_1^s t_M} - \frac{K_p K_v v}{p_1^s - p_2^s} e^{p_2^s t_M}. \qquad (5.6)$$

Namely, this value is the theoretical value of maximum acceleration of the velocity step input of the velocity of equation (5.2) is v. Based on equation (5.6), the theoretical torque output in the bottom of Fig. 5.2 can be calculated.

On the other aspect, the actual system is shown in the top part of Fig. 5.2. It expresses the simple equation including torque saturation, as well as saturation function. However, their meaning are unknown in the actual measurement. Only the input output data can be tested. From the actual measured velocity output of the velocity step input of velocity v in equation (5.2), and the acceleration by calculating difference operation, the torque of actual system is worked out. Through drawing these values and the theoretical acceleration output derived from equation (5.6) respectively by the vertical axis and the horizontal axis of torque saturation. The torque saturation property can be measured. In the next item, the results of measuring concretely the torque saturation curve for an actual mechatronic servo system is illustrated.

5.1.2 Measurement of the Torque Saturation Curve and Experimental Verification

(1) Experimental Equipment

According to the measurement method of the torque saturation property introduced in 5.1.1(3), the torque saturation curve was worked out in the actual mechatronic servo system. The objective is DEC-1 (refer to experiment equipment E.1). The driving and load parts were connected by a rigid body coupling. The experimental conditions are a position loop gain K_p=10[1/s], velocity loop gain K_v=56[1/s] and a sampling time interval Δt_p=10[ms].

(2) Measurement Results of the Torque Saturation Curve

Based on the procedure 5.1.1(3), the measurement of torque saturation curve of DEC-1 was carried out

1. When velocity step input was within the scale of $-13.39 \sim 13.89$[rev/s] (Fig. 5.3(a)) with an interval of 1.39[rev/s], velocity output was read out by the tachogenerator. The typical results with 2.78, 5.56, 8.33, 11.11[rev/s] in Fig. 5.3 were shown. Velocity responses to Fig. (b) respectively with a velocity command for 2.78, 5.56, 8.33, 11.11[rev/s] were shown in Fig. (a). The impact of torque saturation, was that inclination of the response velocity is not bigger than the constant at the start point in Fig. (b), namely, raising part of the velocity with linear inclination, was assured.
2. Through one-order discrete of velocity output, acceleration output (Fig. 5.3(c)) can be calculated.
3. In order to eliminate the noise in the acceleration output wave, an acceleration wave was described by a four-order polynomial $\sum_{i=0}^{4} a_i t^i$ and its coefficient a_i can be worked out by the least-square method (Fig. 5.3(d)). From the peak value of the polynomial function, maximal acceleration can be worked out.
4. From equation (5.6), the peak of acceleration output without torque saturation can be calculated. If K_p=10[1/s], K_v=56[1/s], and t_M of equation (5.5) is 0.04[s], then the peak value of the acceleration output in equation (5.6) can be worked out by $d^2y(0.04)/dt^2 = 7.75v$.
5. The horizontal axis represents the theoretical maximum acceleration output and vertical axis represents actual maximum acceleration output.
6. For a given velocity step and repeating steps 1~5, the torque saturation curve among calculated points can be worked out by interpolating between these points.

According to the above procedure, the results of measuring the torque saturation property are as below and illustrated in Fig. 5.4.

5.1 Measurement Method for the Torque Saturation Property

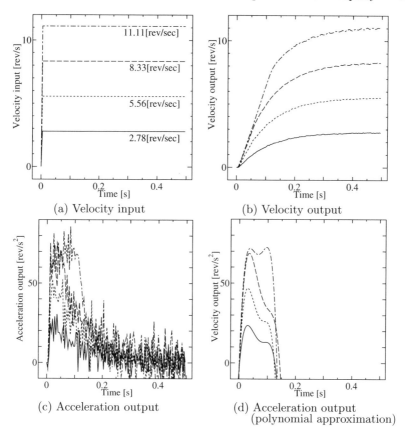

Fig. 5.3. Acceleration output results according to the constant velocity input (2.78, 5.56, 8.33, 11.11[rev/s])

$$sat(x) = \begin{cases} 80 & (80 < x) \\ x & (-80 \leq x \leq 80) \\ -80 & (x < -80). \end{cases} \quad (5.7)$$

From these results, it can be verified that the torque will enter into the saturation region when the command velocity increase and the linear boundary with change of command velocity. In the next part, how much of the actual torque saturation property can be described correctly will be verified.

(3) Verification and Evaluation of the Measured Torque Saturation Property

In order to verify the appropriation of the obtained torque saturation property, the experiment and simulation are compared. In the experiment, an orthogonal type two-axis robot arm is emulated by a combination of two times

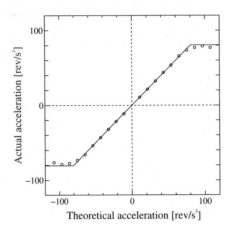

Fig. 5.4. Measurement results of torque saturation property of DEC-1

experiment of DEC-1. In the simulation, each axis is expressed by equation (5.1) and (5.7) and they are combined into two axes. In the experiment and simulation, the objective trajectory of contour control, illustrated in Fig. 5.5, is given as

$$v_x(t) = 9.26 \quad (0 \le t \le 1.08[\text{s}]) \tag{5.8a}$$

$$v_y(t) = \begin{cases} 9.26 & (0 \le t \le 0.54[\text{s}]) \\ -9.26 & (0.54 < t \le 1.08[\text{s}]). \end{cases} \tag{5.8b}$$

In the upper part of Fig. 5.5(a), following the locus of computer simulation results by two methods are shown. In the upper part of Fig. 5.5(b), following locus of experimental results are shown. In Fig. 5.5(a), one of two methods is about simulation without torque saturation. Another is the simulation result including the torque saturation simulated by experiment. The simulation including the torque saturation is performed by replacement of the measured results of the torque saturation curve in Fig. 5.4 with the part without torque saturation (expressed by line), which is in the part of acceleration output of the 2nd order system of the linear model expressed in Fig. 5.2.

In the locus without torque saturation, the corner part indicated by circle has no overshoot. However, in the locus of the system including torque saturation, overshoot exists in the following locus. The simulation results about the latter are almost the same as that in the experimental results shown in Fig. 5.5(b). This property in the time domain can be seen by the position, velocity and acceleration shown from the second to top of Fig. 5.5. Since saturation is not generated owing to no velocity change in the x axis, the impact of saturation is shown in the trajectory of position, velocity, acceleration about features of the y axis. In order to make a comparison for this trajectory in detail, the contour control results of the simulation by using the torque saturation property measured through the proposed procedures are almost the same

5.2 Contour Control Method with Avoidance of Torque Saturation

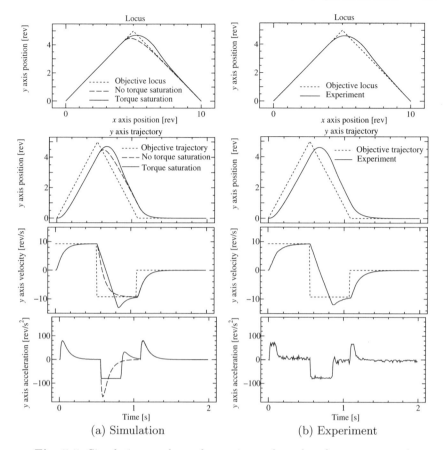

Fig. 5.5. Simulation results and experimental results of contour control

as the results of the DEC-1 experimental results. From the above verification, it shows that the torque saturation curve calculated by the proposed method can match the torque saturation property correctly in the actual system.

5.2 Contour Control Method with Avoidance of Torque Saturation

In order to obtain the high-precision contour control performance in an industrial field and not generate torque saturation, the action is performed with delay totally. Even saturation occurred partly, the working time is lengthened.

In the corner part of the objective locus where occurred torque saturation, the necessary minimal velocity is defined so as to avoid torque saturation. Other parts were moved under designated velocity. This method for minimizing the working time is illustrated. Moreover, the compensation method for dynamic delay in the overall system were also introduced.

The proposed method has a high significance of utilization in the industrial field due to the improvement of control performance without any change of hardware.

5.2.1 Contour Control Performance with Torque Saturation and High-Precision Contour Control Method

(1) Mathematical Model on a Machine Tool

Based on the measured torque saturation in the last section, the contour control method for avoiding torque saturation is introduced in the velocity field when torque saturation occurred. Firstly, the objective is a machine tool, in which each axis can be handled independently in orthogonal coordinates.

The block diagram of a mechatronic servo system, including torque saturation, is shown in Fig. 5.6. In this figure, $U(s)$ and $Y(s)$ are the input and output of a mechatronic servo system, respectively. K_p and K_v have the meanings of K_{p2} and K_{v2} in the middle speed 2nd order model as equation (2.29) of item 2.2.4. The torque saturation property is obviously contained in the servo controller of Fig. 5.6. Dynamics is expressed for each axis independently as (refer to Fig. 5.6)

$$\frac{d^2y(t)}{dt^2} = sat\left[K_v\left\{K_p(u(t) - y(t)) - \frac{dy(t)}{dt}\right\}\right] \tag{5.9}$$

where $sat(z)$ is torque output considering saturation as

$$sat(z) = \begin{cases} A_{\max} & (A_{\max} < z) \\ z & (-A_{\max} \leq z \leq A_{\max}) \\ -A_{\max} & (z < -A_{\max}). \end{cases} \tag{5.10}$$

$A_{\max}$ in equation (5.10) is the maximal acceleration of each axis. $A_{\max}$ is used here to replace the maximum torque in order to have proportional relation with maximum torque.

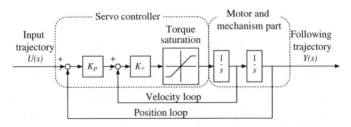

Fig. 5.6. Block diagram of mechatronic servo system including torque saturation

5.2 Contour Control Method with Avoidance of Torque Saturation

In the region without torque saturation, the dynamics of a mechatronic servo system is expressed by the 2nd order system (refer to item 2.2.4) as

$$Y(s) = \frac{K_p K_v}{s^2 + K_v s + K_p K_v} U(s). \tag{5.11}$$

In order to avoid torque saturation, contour control is performed in the region where dynamics of the mechatronic servo system is expressed by equation (5.11).

(2) Contour Control Performance under Torque Saturation

For simplicity, contour control is investigated in an two-dimensional orthogonal coordination. In the three-dimensional case, plate with objective trajectory is considered. The orthogonal coordinate system in the plate is handled similarly for the two-dimensional case.

The objective locus formed by the two lines illustrated by Fig. 5.7, contour control performance is derived under torque saturation. When objective velocity v is constant in contour control, the necessity of realizing the corner part approximated by a polygonal line is that the acceleration should be infinite. Such a polygonal line cannot be described by a mechatronic servo system. In this part, as shown in Fig. 5.7, the objective locus in the corner part approximated by a polygonal line is performed with a circle approximation. The radius of circle r is defined by the working precision ϵ which is the maximum value of errors between the generated locus by circle and polygonal line objective locus. The relationship between circle radius r and the working precision ϵ is as geometry relation in Fig. 5.7 as

$$\cos\left(\frac{\theta_{c2} - \theta_{c1}}{2}\right) = \frac{r}{r + \epsilon}. \tag{5.12}$$

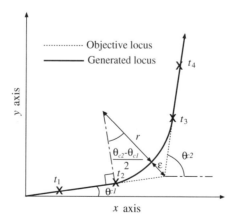

Fig. 5.7. Trajectory generation corresponding to the objective trajectory of two lines

5 Torque Saturation of a Mechatronic Servo System

By solving equation (5.12) on r, circle radius r satisfying working precision ϵ can be calculated as

$$r = \frac{\epsilon \cos\{(\theta_{c2} - \theta_{c1})/2\}}{1 - \cos\{(\theta_{c2} - \theta_{c1})/2\}}. \tag{5.13}$$

The maximal acceleration $A_{\max}$ obtained from the mechatronic servo system is dependent on the equipment. As shown in Fig. 5.7, the locus angle θ_{c1} and θ_{c2} of the objective locus are given. According to the working task of contour control, minimal velocity $v_{\min}$ needed with the lowest limitation in operation under objective velocity v and working precision ϵ are given. In the following part, equations related with minimal velocity $v_{\min}$ in the objective of control performance and maximal acceleration $A_{\max}$, which is about relationship between working precision ϵ and torque, are derived.

(i) *Boundary values of working precision ϵ with a given maximal velocity $v_{\min}$*

Acceleration of the circle trajectory is $a = r\omega^2$. If $\omega = v/r$, then $r = v^2/a$. Therefore, the region of circle radius with minimal velocity $v_{\min}$ and maximal velocity $A_{\max}$ is expressed as

$$r \geq \frac{v_{\min}^2}{A_{\max}}. \tag{5.14}$$

The relationship between radius r and the working precision ϵ is expressed by the equation (5.13) and the solution of ϵ is as

$$\epsilon = \left[\frac{1}{\cos\{(\theta_{c2} - \theta_{c1})/2\}} - 1\right] r.$$

From the relationship between this equation and equation (5.14), the scale of working precision ϵ, if given minimal velocity $v_{\min}$, is as

$$\epsilon \geq \left[\frac{1}{\cos\{(\theta_{c2} - \theta_{c1})/2\}} - 1\right] \frac{v_{\min}^2}{A_{\max}}. \tag{5.15}$$

(ii) *Boundary values of minimal $v_{\min}$ when given working precision ϵ*

Radius r is expressed by equation (5.13) when given working precision ϵ. The minimal velocity $v_{\min}$ with this radius r and maximal acceleration $A_{\max}$ can be expressed as $v_{\min} \leq \sqrt{A_{\max} r}$ from equation (5.14). If we put r of this equation into equation (5.13), the scale of minimal velocity $v_{\min}$, if given working precision ϵ, is expressed as

$$v_{\min} \leq \sqrt{\frac{A_{\max} \epsilon \cos\{(\theta_{c2} - \theta_{c1})/2\}}{1 - \cos\{(\theta_{c2} - \theta_{c1})/2\}}}. \tag{5.16}$$

5.2 Contour Control Method with Avoidance of Torque Saturation

(3) Contour Control Considering Torque Saturation

In the contour control of an industrial mechatronic servo system, motion is performed in the region without generating torque saturation. In order to implement it, the trajectory of mechatronic servo system should be determined without torque saturation. Fig. 5.8 illustrates the contour control structure of a mechatronic servo system. The contour control considering torque saturation is divided into two big parts. One is the generation part of the trajectory in working coordinates without torque saturation. Another is the compensation part of dynamics of the mechatronic servo system.

For generation of trajectory $(w_x(t), w_y(t))$, a locus is generated by satisfying the working precision ϵ between the objective locus (r_x, r_y) and the generated locus (w_x, w_y) without torque saturation in a mechatronic servo system as shown in Fig. 5.8 firstly. The velocity given in locus (w_x, w_y) generation is approximated with the objective velocity v with a limitation in the region without torque saturation.

If directly using the generated trajectory $(w_x(t), w_y(t))$ as an input trajectory $(u_x(t), u_y(t))$, following the locus (x, y) generated from the locus (w_x, w_y) will be degraded because of the dynamics of the mechatronic servo system. If using the inverse dynamics of the mechatronic servo system in equation (5.11) without torque saturation, the input trajectory $(u_x(t), u_y(t))$ can be adopted with revised generated trajectory $(w_x(t), w_y(t))$. Then, any delay of the mechatronic servo system is compensated, and the following trajectory $(p_x(t), p_y(t))$ is consistent with the generated trajectory $(w_x(t), w_y(t))$. Moreover, the following locus (x, y) is satisfied with working precision of ϵ.

(4) Trajectory Generation Considering Torque Saturation

For an objective locus (r_x, r_y) generated from two lines for approximating the trajectory shown in Fig. 5.7, the trajectory generation method, if generating a trajectory along the time shift under the limitation of the torque of the mechatronic servo system, is explained below.

1. When there exists an angle in the objective locus (r_x, r_y), the angle will be approximated by a circle satisfying working precision ϵ.
2. Radius r of the circle included in the locus (w_x, w_y) is calculated by a tangent velocity between the minimal radius $r_{\min}(= v^2/A_{\max})$ satisfying torque constraints and the maximal acceleration $A_{\max}$.

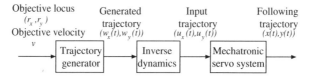

Fig. 5.8. Contour control structure of mechatronic servo system including torque saturation

112   5 Torque Saturation of a Mechatronic Servo System

a) If $r \geq r_{\min}$: generated trajectory $(w_x(t), w_y(t))$ is calculated for changing the objective tangent velocity into tangent velocity.
b) If $r < r_{\min}$: trajectory is generated according to the following procedure.
   i. In the region from t_1 to t_2, the tangent velocity is decelerated with maximal deceleration of $-A_{\max}$ from v to $v_{\min}$ (the tangent velocity is $v_{\min} = \sqrt{A_{\max} r}$ if the acceleration of radius r circle is $A_{\max}$).
   ii. In the region from t_2 to t_3, the locus is described by circle.
   iii. In the region from t_3 to t_4, the tangent velocity is accelerated with a maximal acceleration of $A_{\max}$ from $v_{\min}$ to v.
3. In the beginning point and end point of the objective locus, acceleration and deceleration are performed with a maximal acceleration $A_{\max}$.

Based on the above introduced procedure, a trajectory $(w_x(t), w_y(t))$ can be generated without torque saturation and the generated locus (w_x, w_y) can be made consistent with the objective locus (r_x, r_y) within the working precision ϵ.

In the case of 2b, trajectory generation can be derived by

$$w_x(t) = \begin{cases} vt \cos\theta_{c1} & (t \leq t_1) \\ w_x(t_1) + \left\{ v(t-t_1) - \dfrac{A_{\max}(t-t_1)^2}{2} \right\} \cos\theta_{c1} & (t_1 < t \leq t_2) \\ w_x(t_2) + r\left[\sin\left\{\theta_{c1} + \dfrac{v_{\min}(t-t_2)}{r}\right\} - \sin\theta_{c1} \right] & (t_2 < t \leq t_3) \\ w_x(t_3) + \left\{ v(t-t_3) + \dfrac{A_{\max}(t-t_3)^2}{2} \right\} \cos\theta_{c2} & (t_3 < t \leq t_4) \\ w_x(t_4) + vt\cos\theta_{c2} & (t_4 < t) \end{cases} \quad (5.17a)$$

$$w_y(t) = \begin{cases} vt \sin\theta_{c1} & (t \leq t_1) \\ w_y(t_1) + \left\{ v(t-t_1) - \dfrac{A_{\max}(t-t_1)^2}{2} \right\} \sin\theta_{c1} & (t_1 < t \leq t_2) \\ w_y(t_2) + r\left[\cos\left\{\theta_{c1} + \dfrac{v_{\min}(t-t_2)}{r}\right\} - \cos\theta_{c1} \right] & (t_2 < t \leq t_3) \\ w_y(t_3) + \left\{ v(t-t_3) + \dfrac{A_{\max}(t-t_3)^2}{2} \right\} \sin\theta_{c2} & (t_3 < t \leq t_4) \\ w_y(t_4) + vt\sin\theta_{c2} & (t_4 < t) \end{cases} \quad (5.17b)$$

where the time interval of deceleration and acceleration is $t_4 - t_3 = t_2 - t_1 = (v - v_{\min})/A_{\max}$, describing the time of the circle is $t_3 - t_2 = r(\theta_{c2} - \theta_{c1})/v_{\min}$. This method is performed under condition of 2b $r < r_{\min}$ and with the lowest

5.2 Contour Control Method with Avoidance of Torque Saturation

limitation of velocity for preventing a rapid change in velocity. Besides, the control time becomes longer in order to describe a circle. The high-precision contour control will be performed under the conditions of that following the locus (x, y) at the angle part also should be satisfied torque constraints, and the generated locus (w_x, w_y) should be in agreement with the objective locus (r_x, r_y) within the working precision ϵ.

(5) Delay Compensation Based on Inverse Dynamics

In order to compensate for the dynamics of the mechatronic servo system, the trajectory should be revised by using inverse dynamics. Although the inverse dynamics of equation (5.11) contains a second-order differential, the trajectory $(w_x(t), w_y(t))$ is possible to obtain a 2nd order differential, compensation based on inverse dynamics can be realized to design acceleration without torque saturation. The inverse dynamics of a mechatronic servo system as in equation (5.11) without torque saturation is expressed as

$$F(s) = \frac{s^2 + K_v s + K_p K_v}{K_p K_v}. \tag{5.18}$$

The input trajectory $(u_x(t), u_y(t))$ is derived according to a revised trajectory $(w_x(t), w_y(t))$ based on inverse dynamics (5.18) as

$$u_x(t) = w_x(t) + \frac{1}{K_p}\frac{dw_x(t)}{dt} + \frac{1}{K_p K_v}\frac{d^2 w_x(t)}{dt^2} \tag{5.19a}$$

$$u_y(t) = w_y(t) + \frac{1}{K_p}\frac{dw_y(t)}{dt} + \frac{1}{K_p K_v}\frac{d^2 w_y(t)}{dt^2}. \tag{5.19b}$$

When input trajectory $(u_x(t), u_y(t))$ are adopted as the command of the mechatronic servo system, the following trajectory $(p_x(t), p_y(t))$ can be in good agreement with the generated trajectory $(w_x(t), w_y(t))$.

(6) Contour Control Algorithm Considering Torque Saturation

The procedure of contour control considering torque saturation is illustrated as below.

1. A trajectory is generated based on equation (5.17a), (5.17b) according to the procedure of 5.2.1(4) from the objective trajectory $(r_{xi}(t), r_{yi}(t))$.
2. An input trajectory is calculated for compensating delay of dynamics by using inverse dynamics of equation (5.19)
3. Input command of objective trajectory, which can compensate for the dynamics delay of the mechatronic servo system, is given.

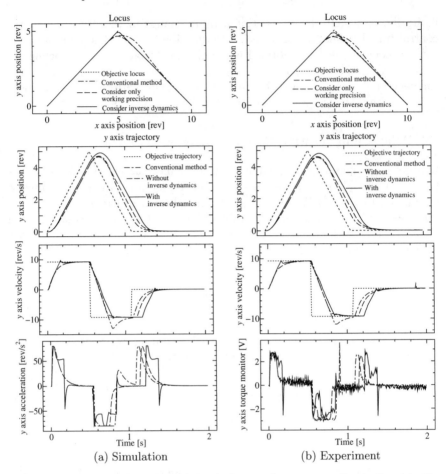

Fig. 5.9. Experimental results and simulation results corresponding to the objective trajectory of two lines

5.2.2 Experimental Verification of Contour Control Considering Torque Saturation

(1) Experiment Using DEC-1

In order to verify the effectiveness of the contour control method avoiding torque saturation, a computer simulation and experiment using the DEC-1 (experiment equipment referring E.1) were carried out. As contour control approaches, three methods are compared, i.e., conventional method with original objective trajectory usually used in the industrial field, considering only working precision without performing acceleration and deceleration, and contour control avoiding torque saturation. The conditions of computer simulation and experiment are as below: position loop gain $K_p = 10[1/s]$, velocity

5.2 Contour Control Method with Avoidance of Torque Saturation

loop gain $K_v = 56[1/\text{s}]$, maximal acceleration $A_{\max} = 80[\text{rev}/\text{s}^2]$, sampling time interval 10[ms], working precision $\epsilon = 0.1[\text{rev}]$, objective tangent velocity $v = 13.1[\text{rev}/\text{s}]$. The objective trajectory is given as

$$\frac{dr_x(t)}{dt} = 9.26 \qquad (0 \le t \le 1.08[\text{s}]) \tag{5.20a}$$

$$\frac{dr_y(t)}{dt} = \begin{cases} 9.26 & (0 \le t \le 0.54[\text{s}]) \\ -9.26 & (0.54 < t \le 1.08[\text{s}]). \end{cases} \tag{5.20b}$$

Input trajectory $(u_x(t), u_y(t))$ is derived according to the procedure of 5.2.1(4). In Fig. 5.9, the computer simulation results and experimental results are illustrated. The acceleration output in the experimental results is measured by a torque monitor. As shown in Fig. 5.9, the following locus generated overshoot is based on the conventional method. This overshoot is not permitted to occur in contour control in industry (refer to 1.1.2 item 3). However, overshoot does not occur in the proposed method which considers working precision. In addition, the following locus has a large error compared with objective locus in the conventional method, but in the proposed method, the following locus is almost the same as the objective locus when considering working precision. In the experimental results, the locus error is 0.17[rev]. From the acceleration output in the experimental results shown in the figure, torque saturation is generated. The torque saturation is 3[V] response of the torque monitor. Concerning the bad impact of the conventional method, the tangent velocity by conventional method will become larger than the objective tangent velocity $v = -9.26[\text{rev}/\text{s}]$. At the peek point, the velocity is $-11.5[\text{rev}/\text{s}]$ in the simulation and $-11.0[\text{rev}/\text{s}]$ in the experimental results. However, in the contour control method avoiding torque saturation, the tangent velocity is also consistent with the objective tangent velocity. From these results, the proposed method is effective in comparing other two methods.

(2) Experiment Using an Articulated Robot Arm (Performer MK3S)

The proposed contour control method considering torque saturation was adopted for an articulated robot arm (Performer MK3S; experiment device refers to E.3). There are nonlinear transforms between working coordinates and joint coordinates adopted in the articulated robot arm. As introduced above, the contour control method avoiding torque saturation cannot be adopted without change. If generating trajectory considering torque saturation in working coordinates and compensating for delay in joint coordinates, the proposed method can be adopted. In the delay compensation in joint coordinates, modified taught data method (refer to section 6.1) is used here. Besides, the relationship between maximal acceleration $a_{\max}$ in joint coordinates and maximal acceleration $A_{\max}$ in working coordinates is calculated according to coordinate transform by using Jacobian with a reference input time interval.

Although Performer MK3S uses 5 axes for a 5-freedom-degree articulated robot arm, only two axes are used in the experiment. The servo motor in each axis is connected with the servo controller for carrying out velocity and current control. The servo controller is connected with the computer when performing position control. In each axis, an AC servo motor (rated speed 3000[rpm]) is used and driving arm through deceleration device. The conditions of the device are: position loop gain $K_p = 25[1/s]$, velocity loop gain $K_v = 150[1/s]$, maximum acceleration $a_{max} = 11.0[\text{rad}/s^2]$, sampling time interval $\Delta t = 6[\text{ms}]$(refer to section 3.1), length of arm $l_1 = 0.25[\text{m}]$, $l_2 = 0.215[\text{m}]$, gear ratio of each axis $n_1 = 160$, $n_2 = 161$. In the experiment, the value multiplying position loop gain K_p in the error between position input and motor position output are put into the motor as velocity input through a D/A converter.

(i) Supposed torque saturation generation

The Performer MK3S used in the experiment can output very large amounts of torque. In order to verify the significance of the proposed method, the supposed torque saturation can be generated by this device. This method focuses on velocity input. If the actual measured angular acceleration output multiplying velocity loop gain K_v with the error between velocity input v_i and output v_f satisfied

$$|K_v(v_i - v_f)| > a_{max} \tag{5.21}$$

velocity input v_i is changed as

$$v_i = \text{sign}(v_i - v_f)\left(\frac{a_{max}}{K_v}\right) + v_f \tag{5.22}$$

angular acceleration is not over a_{max}. Torque saturation is changeable depended on the device type. Based on the proposed method, the experiment is realized in the same device considering various torque properties.

(ii) Simulation and experimental results

Fig. 5.10 illustrates the locus for four methods in 5.11, synthesized velocity and simulation results and experimental results of the B axis acceleration with saturation. (a) conventional method (objective trajectory is used as input of the robot arm without any change), (b)conventional method in the state with supposed torque saturation generation, (c) contour control method (considering precision) considered torque saturation, (d) contour control method (considering velocity) considered torque saturation are adopted. The conditions of the simulation are designated tangent velocity $v = 0.15[\text{m/s}]$, objective locus $0.05[\text{m}]$ length two lines of $(0.135, 0.365) \sim (0.185, 0.365) \sim (0.185, 0.415)$ which is turned as a vertical angle. As introduced in 5.2.1(4), maximal acceleration a_{max} in joint coordinates and maximal acceleration in working coordinates given from the objective are calculated as $A_{max} = 1.0[\text{m}/s^2]$. In

5.2 Contour Control Method with Avoidance of Torque Saturation

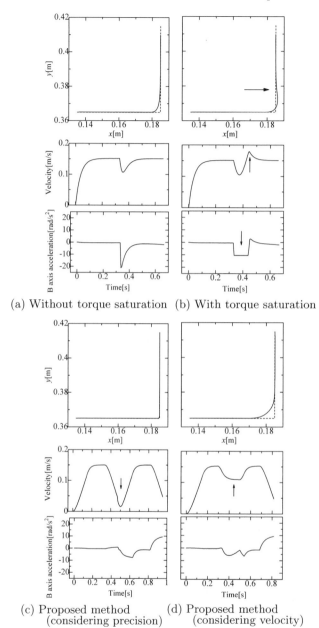

(a) Without torque saturation (b) With torque saturation

(c) Proposed method (considering precision) (d) Proposed method (considering velocity)

Fig. 5.10. Simulation results

5 Torque Saturation of a Mechatronic Servo System

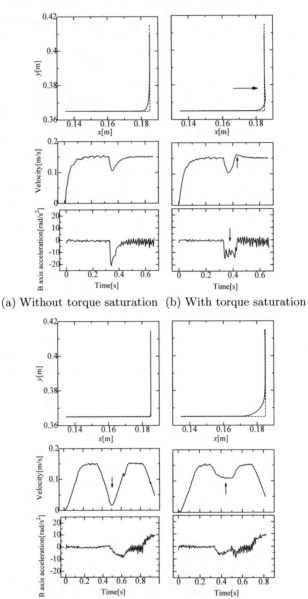

(a) Without torque saturation (b) With torque saturation

(c) Proposed method (considering precision) (d) Proposed method (considering velocity)

Fig. 5.11. Experimental results

5.2 Contour Control Method with Avoidance of Torque Saturation

the contour control considering torque saturation for focusing on precision in Fig. (c), the working precision is $\epsilon = 1.0 \times 10^{-4}$[m] focusing on locus, minimal velocity $v_{\min} = 0.0155$[m/s] is given when velocity is decreased to 10% of objective velocity. In the contour control considering torque saturation for focusing on velocity in Fig. (d), there exists a decrease of contour control precision when the response cannot be fit for the situation that velocity is over dropped at the corner at the operation of laser cutting, or input current of laser is over reduced, or increasing current cost so much time at velocity increasing. For these cases, the delay issue of input current response will disappear when velocity is only equal to 70% of the objective velocity. Then, the working precision was calculated under the condition that the velocity was decreased till 70% of objective velocity. If $\epsilon = 0.005$[m], minimal velocity $v_{\min} = 0.1$[m/s] is given when velocity is decreased to 70% of objective velocity. The common pole of regulator in Fig. (c) and (d) was given as $\gamma = -30$.

In the following locus of Fig. (a), the deterioration of locus as roundness at the corner part of the simulation and experiment can be found. The reason for deterioration is the delay dynamics of the robot arm and it can be understood even from the results of acceleration to be not linked to torque saturation. On the other hand, the marked part of B axis acceleration exist 0.33~0.44[s] saturation by observing the results of each axis acceleration in the simulation and experimental results in Fig. (b). In addition, the error in the experiment is smaller in the simulation results and experimental results. With same trend at the marked part of the following locus, in the simulation error is 1.35[mm], but in the experiment is 0.74[mm]. At the marked combined velocity, in the simulation the overshoot is 0.3[m/s], but in the experiment is 0.12[m/s]. Overshoot must be avoided as much as possible in order to improve precision (refer to 1.1.2 item 3). From the simulation and experimental results in Fig. (c), there are no overshoots in the following locus results. From the combined velocity, spending more time than Fig. (a) and (b) at the marked corner part for using necessary minimal velocity. Hence, the dynamics of the robot arm is compensated and there is no torque limitation. In addition, the minimal velocity is satisfied as $v_{\min} = 0.015$[m/s]. From the simulation and experimental results in Fig. (d), there is no overshoot in the following locus results, and the designated working precision is satisfied as $\epsilon = 0.005$[m]. Spending time is not longer than Fig. (a), (b), and there is no torque limitation. Additionally, from the synthesis velocity, minimal velocity is larger than $v_{\min} = 0.1$[m/s] in order to reduce the velocity at the marked corner part.

From the above simulation and experimental results, the contour control method considering torque saturation satisfies working precision and minimal velocity within the torque saturation, and it can be realized within the limitation of contour control performance.

6
The Modified Taught Data Method

In order to realize the movement of an industrial robot, the given objective trajectory is always used without any change when their coordinate values which are the taught data obtained from the teaching. Therefore, in the movement response of the robot at the playback, the errors between the objective locus and the following locus of the robot appeared because of the time delay generated at each axis. In this chapter, the modified taught data method is proposed in order to improve the precision of the trajectory in the contour control.

6.1 Modified Taught Data Method Using a Mathematical Model

In the operation of the robot, the practician, who is performing the teaching of the robot in the industrial field, improved the precision of the contour control of the robot successfully through the teaching points with a little over movement from the actual objective points at the corner part of objective locus (modified taught data). However, this method can be only adapted for the limited action situation.

From the investigation of the adopted method by the practician and the reasons of performance improvement, the deterioration of control performance owing to the dynamics delay of the mechatronic servo system and the realization method of dynamic compensation (modified taught data) have been found. With the model of a mechatronic servo system in chapter 1, the modified taught data method with pole assignment regulator for the dynamic compensation was proposed and the construction of the modification element was introduced. In order to use this method for the semi-closed pattern which is without a sensor for measuring the tip position of the robot arm in the mechatronic servo system (refer to 1.1.2 item 5), the modification element was revised from the closed-loop form with the control law to the open-loop form as (6.7), (6.25). With the characteristic evaluation of the obtained modification element of the taught data by a frequency transfer function, the realization of the phase-lead compensation was known.

According to this modified taught data method, any shape of the objective locus not only the rectangle can be realized. If the servo parameters K_p, K_v were clear and understood, this method can be adopted for any mechatronic servo system. Also, it is only necessary to revise the software in this method. The existing hardware does not need to be changed. Therefore, this method is very useful in the industrial field.

6.1.1 Derivation of the Modified Taught Data Method

(1) Concept of the Modified Taught Data Method

In the working coordinates of a mechatronic servo system, the relationship between the input and output of the each independent coordinate axis can be expressed independently as

$$Y(s) = G(s)U(s) \tag{6.1}$$

where $U(s)$ denotes taught data, $Y(s)$ the following trajectory of the mechatronic servo system and $G(s)$ the dynamics of the mechatronic servo system. The teaching playback robot refers to the semi-closed type control system (refer to 1.1.2 item 5) with the feedforward control, but without the measure of the tip position or velocity of the mechatronic servo system and the change of hardware. Moreover, the modification element $F(s)$ for the objective trajectory $R(s)$ through the taught data $U(s)$ can be generated. That means that the taught data $U(s)$ can be expressed as,

$$U(s) = F(s)R(s). \tag{6.2}$$

Fig. 6.1 shows the block diagram of the modified taught data method. In order to realize the desired control performance $Y(s) = R(s)$, i.e., keeping the mechatronic servo system the concordance with objective trajectory, the modification element $F(s)$ was required for the inverse dynamics $G^{-1}(s)$ of the mechatronic servo system. However, in the design of the modification element by $F(s) = G^{-1}(s)$, if there is no proper inverse dynamic $G^{-1}(s)$, the taught data will diverge when the objective trajectory is not differential. Therefore,

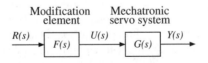

Fig. 6.1. Block diagram of the modified taught data method

6.1 Modified Taught Data Method Using a Mathematical Model

the mechatronic servo system will be expressed by the state-space representation and the modification element will be designed with the pole assignment regulator (refer to appendix A.3) in order to change the mechatronic servo system $F(s)G(s)$ into the appropriate closed-loop control system.

(2) A modified taught data method based on the 1st order model

(i) Mathematical model

Firstly, deriving the modification element easily, the 1st order model of the mechatronic servo system is derived by the modified taught data method. When the actuator of the mechatronic servo system, i.e., the velocity of the servo motor, is moved under 1/20 of the rated value, the whole control system of the mechatronic servo system including control equipment, servo system and mechanism shown in Fig. 6.2 can be expressed as the 1st order model in the working coordinates with each independent coordinate axis (refer to the 2.2.3)

$$G_1(s) = \frac{K_p}{s + K_p} \tag{6.3}$$

where K_p denotes the meaning of K_{p1} in the equation (2.23) of the low speed 1st order model of 2.2.3.

(ii) Modification element

As expressed in the state space of the equation (6.3) of the mechatronic servo system, the modification element $F_1(s)$ is derived by the pole assignment regulator (refer to the appendix A.3). For the objective trajectory $r(t)$, assume $dr(t)/dt \simeq 0$. From the equation (6.3) and the assumption $dr(t)/dt \simeq 0$, the mechatronic servo system expressed by a state-space representation is changed as

$$\frac{dx(t)}{dt} = -K_p x(t) + K_p r^*(t) \tag{6.4}$$
$$x(t) = y(t) - r(t)$$
$$r^*(t) = u(t) - r(t).$$

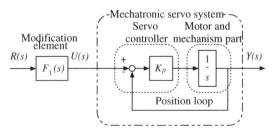

Fig. 6.2. Block diagram of the modified taught data method based on the 1st order model

If the state equation (6.4) can be derived with the assumption $dr(t)/dt \simeq 0$, the pole assignment regulator can be adopted and the modification term of the taught data $r^*(t)$ is easily derived. Thus, as one of the key conditions the meaning of the assumption $dr(t)/dt \simeq 0$ which brought about good results will be explained in the 6.1.2.

From the pole assignment regulator, the control input is given as

$$r^*(t) = K_s x(t) \tag{6.5}$$

where K_s denotes the feedback gain of the regulator. The relationship between the feedback gain K_s and the pole of the regulator γ is shown as

$$\gamma = -K_p(1 - K_s). \tag{6.6}$$

When the equation (6.5) and equation (6.6) are input into the equation (6.3) which expresses the 1st order model, the taught data $u(t)$ is given as

$$\frac{du(t)}{dt} - \gamma u(t) = -\frac{\gamma}{K_p}\left(\frac{dr(t)}{dt} + K_p r(t)\right). \tag{6.7}$$

From the Laplace transform of the equation (6.7)(refer to the appendix A.1), the modification element $F_1(s)$ is given as

$$F_1(s) = -\frac{\gamma(s + K_p)}{K_p(s - \gamma)}. \tag{6.8}$$

From the solution of the differential equation (6.7) about $u(t)$, the taught data $u(t)$ can be calculated based on the 1st order model of the mechatronic servo system.

When the modification element $F_1(s)$ is adopted in the mechatronic servo system $G_1(s)$, the control system of the robot arm after revision can be changed as

$$Y(s) = \frac{-\gamma}{s - \gamma} R(s). \tag{6.9}$$

From the comparison between the original mechatronic servo system (6.3) and the revised mechatronic servo system (6.9), the modification element changes the pole of the mechatronic servo system from $-K_p$ to γ.

(iii) Selection of the pole

The selection of the regulator pole γ is given by the designer in the equation (6.8) of the modification element is introduced. Firstly, in order to improve the control performance of the mechatronic servo system, it is necessary to satisfy the following equation so that the response of the control system of the mechatronic servo system after revision is faster than that before revision.

$$\gamma \leq -K_p. \tag{6.10}$$

Then, the velocity limitation of the servo motor, i.e., the actuator of the mechatronic servo system, must be considered when using the modified taught data method in the actual mechatronic servo system. When the maximum velocity of the servo motor is $V_{\max}$, the velocity limitation is shown as,

$$|K_p\{u(t) - y(t)\}| \leq V_{\max}. \tag{6.11}$$

The left-hand side of (6.11) denotes the velocity input of the servo motor. In fact, the computer simulation of the modified taught data method with the pole which has a certain error in the left side of (6.11) is made. The minimal pole which is satisfied the by conditions of (6.10) and (6.11) is selected.

(3) Modified Taught Data Method Based on the 2nd Order Model

(i) Mathematical model

When the velocity of the motion of the mechatronic servo system becomes high and the velocity of the servo motor is between $1/5 \sim 1/20$ of the rated value, considering the characteristics of the velocity control of the servo motor and the control system of the whole mechatronic servo system shown in Fig. 6.3, it is necessary to express each coordinate independently with the 2nd order model as (refer to 2.2.4)

$$G_2(s) = \frac{K_p K_v}{s^2 + K_v s + K_p K_v} \tag{6.12}$$

where K_p, K_v have the meanings of K_{p2}, K_{v2} in (2.29) of the middle speed 2nd order model in 2.2.4, respectively.

(ii) Modification element

The mechatronic servo system is expressed by a state-space representation based on the 2nd order model (6.12). The modification element can be derived by the pole assignment regulator (refer to appendix A.3) and the minimum

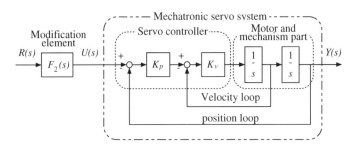

Fig. 6.3. Block diagram of the modified taught data method based on the 2nd order model

order observer (refer to appendix A.4). Since the 2nd order model contains the velocity loop, the derivation of the modification element $F_2(s)$ is more complex than that of the 1st order model.

From (6.12) and the assumption of $d^2r(t)/dt^2 + K_v dr(t)/dt \simeq 0$, the mechatronic servo system can be expressed with a state-space representation as,

$$\frac{d\boldsymbol{x}(t)}{dt} = A\boldsymbol{x}(t) + br^*(t), \quad y^*(t) = c\boldsymbol{x}(t) \tag{6.13}$$

$$A = \begin{pmatrix} -K_v & 1 \\ -K_p K_v & 0 \end{pmatrix}, \quad b = \begin{pmatrix} 0 \\ K_p K_v \end{pmatrix}, \quad c = (1 \ 0) \tag{6.14a}$$

$$\boldsymbol{x}(t) = \begin{pmatrix} y^*(t) \\ dy^*(t)/dt + K_v y^*(t) \end{pmatrix} \tag{6.14b}$$

$$y^*(t) = y(t) - r(t), \quad r^*(t) = u(t) - r(t). \tag{6.14c}$$

The assumption $d^2r(t)/dt^2 + K_v dr(t)/dt \simeq 0$ is adapted for regulator theory. The significance of introducing the assumption $d^2r(t)/dt^2 + K_v dr(t)/dt \simeq 0$ is explained in 6.1.2.

The state-space representation equation (6.13) of the mechatronic servo system is fit for the pole assignment regulator and a minimum order observer. The pole assignment regulator is expressed as (refer to the appendix A.3)

$$r^*(t) = (f_1 \ f_2)\hat{\boldsymbol{x}}(t) \tag{6.15}$$

$$f_1 = 1 - \frac{K_v}{K_p} - \frac{\gamma_1 + \gamma_2}{K_p} - \frac{\gamma_1 \gamma_2}{K_p K_v}$$

$$f_2 = \frac{1}{K_p} + \frac{\gamma_1 + \gamma_2}{K_p K_v}.$$

Moreover, the minimum order observer is changed as (refer to the appendix A.4)

$$\frac{dz(t)}{dt} = \mu z(t) - (K_p K_v + \mu K_v + \mu^2)y^*(t) + K_p K_v r^*(t) \tag{6.16a}$$

$$\hat{\boldsymbol{x}}(t) = \begin{pmatrix} 0 \\ 1 \end{pmatrix} z(t) + \begin{pmatrix} 1 \\ -\mu \end{pmatrix} y^*(t). \tag{6.16b}$$

When we input $\hat{\boldsymbol{x}}(t)$ into (6.15), the control input $r^*(t)$ can be derived as

$$r^*(t) = (f_1 - \mu f_2)y^*(t) + f_2 z(t). \tag{6.17}$$

In order to obtain the modification element $F_2(s)$, (6.14c), (6.16a) and (6.17) are transformed into the frequency domain as

6.1 Modified Taught Data Method Using a Mathematical Model

$$Y^*(s) = Y(s) - R(s), \quad R^*(s) = U(s) - R(s) \tag{6.18a}$$

$$Z(s) = \frac{-K_p K_v - \mu K_v - \mu^2}{s - \mu} Y^*(s) + \frac{K_p K_v}{s - \mu} R^*(s) \tag{6.18b}$$

$$R^*(s) = (f_1 - \mu f_2) Y^*(s) + f_2 Z(s). \tag{6.18c}$$

When we input (6.18b) into (6.18c), the relationship between $R^*(s)$ and $Y^*(s)$ can be obtained as

$$R^*(s) = \frac{(s - \mu) f_1 - (\mu s + K_p K_v + \mu K_v) f_2}{s - \mu - f_2 K_p K_v} Y^*(s). \tag{6.19}$$

From (6.18a) and (6.19), $U(s)$ can be given with $R(s)$ and $Y(s)$

$$U(s) = \{1 - P(s)\} R(s) + P(s) Y(s) \tag{6.20}$$

where

$$P(s) = \frac{(s - \mu) f_1 - (\mu s + K_p K_v + \mu K_v) f_2}{s - \mu - f_2 K_p K_v}. \tag{6.21}$$

The relationship between the objective trajectory $R(s)$ and the following trajectory of the mechatronic servo system $Y(s)$ is changed from (6.12) and (6.20) as

$$Y(s) = \frac{G_2(s)\{1 - P(s)\}}{1 - G_2(s) P(s)} R(s). \tag{6.22}$$

Finally, the modification element $F_2(s)$ is derived from (6.22) as

$$F_2(s) = \frac{1 - P(s)}{1 - G_2(s) P(s)}. \tag{6.23}$$

When we input f_1 and f_2, the modification element $F_2(s)$ can be expressed by the poles of the regulator $\gamma_1, \gamma_2 (< 0)$, the pole of the observer $\mu (< 0)$ and the servo parameter K_p, K_v as

$$F_2(s) = \frac{\alpha_3 s^3 + \alpha_2 s^2 + \alpha_1 s + \alpha_0}{(s - \gamma_1)(s - \gamma_2)(s - \mu)} \tag{6.24}$$

$$\alpha_0 = -\mu \gamma_1 \gamma_2$$

$$\alpha_1 = (K_v + \mu)(\gamma_1 + \gamma_2) + K_v^2 + \gamma_1 \gamma_2 + K_v \mu - \frac{\mu \gamma_1 \gamma_2}{K_p}$$

$$\alpha_2 = \frac{1}{K_p} \{(K_v + \mu)(\gamma_1 + \gamma_2) + K_v^2 + \gamma_1 \gamma_2 + K_v \mu\} - \frac{\mu \gamma_1 \gamma_2}{K_p K_v}$$

$$\alpha_3 = \frac{1}{K_p K_v} \{(K_v + \mu)(\gamma_1 + \gamma_2) + K_v^2 + \gamma_1 \gamma_2 + K_v \mu\}.$$

In the time domain, the modification element $F_2(s)$ can be transformed as

$$\left(\frac{d}{dt} - \gamma_1\right)\left(\frac{d}{dt} - \gamma_2\right)\left(\frac{d}{dt} - \mu\right) u(t)$$
$$= \left(\alpha_3 \frac{d^3 r(t)}{dt^3} + \alpha_2 \frac{d^2 r(t)}{dt^2} + \alpha_1 \frac{dr(t)}{dt} + \alpha_0 r(t)\right). \tag{6.25}$$

According to the solution of the differential equation (6.25) about $u(t)$, the modified taught data $u(t)$ can be calculated based on the 2nd order model.

From the modification element $F_2(s)$ and the mechatronic servo system (6.12), the mechatronic servo system after revision can be described as

$$Y(s) = \frac{\beta_1 s + \beta_0}{(s - \gamma_1)(s - \gamma_2)(s - \mu)} R(s) \qquad (6.26)$$

$$\beta_0 = -\mu \gamma_1 \gamma_2$$

$$\beta_1 = (K_v + \gamma_1 + \gamma_2)(K_v + \mu) + \gamma_1 \gamma_2.$$

(iii) Selection of a pole

In the design of the modification element as (6.24), the appropriate selection poles of the regulator γ_1, γ_2 and the pole of the observer is necessary. Since the pole of the observer should be smaller than the pole of the regulator, i.e.,

$$\mu < \min(\gamma_1, \gamma_2). \qquad (6.27)$$

concerning the pole of the regulator, $\gamma_1 \leq \gamma_2$ is assumed without losing generality. If applying the modified taught data method in the actual mechatronic servo system, the overshoot must be avoided in the following trajectory of the mechatronic servo system (refer to 1.1.2 item 3). In the third order system (6.26) with one zero, the condition of not generating an overshoot is that it is better to define the most pole below the zero. Therefore, the pole of the regulator is selected for meeting the following condition,

$$\gamma_2 \geq \frac{\mu \gamma_1 \gamma_2}{(K_v + \gamma_1 + \gamma_2)(K_v + \mu) + \gamma_1 \gamma_2}. \qquad (6.28)$$

With the transformation of (6.28) as

$$(K_v + \gamma_2)(K_v + \mu + \gamma_1) \geq 0 \qquad (6.29)$$

because of the $\mu < \gamma_1 \leq \gamma_2 < 0$, it can be obtained as

$$\gamma_2 \geq -K_v \qquad (6.30)$$

In order to realize the fastest response of the condition (6.30), the pole is as $\gamma_2 = -K_v$ and defining

$$Y(s) = \frac{\mu \gamma_1}{(s - \gamma_1)(s - \mu)} R(s). \qquad (6.31)$$

From the original mechatronic servo system (6.12) and the mechatronic servo system after revision (6.31), the modification element transforms the poles of the mechatronic servo system from $(-K_v \pm \sqrt{K_v^2 - 4K_v K_p})/2$ to γ_1 and μ. Similar as the 1st order system, since the control system of mechatronic servo system after revision becomes faster than that before revision in order to improve the control performance of the mechatronic servo system, γ_1 should be satisfied

$$\gamma_1 \leq \frac{-K_v - \sqrt{K_v^2 - 4K_v K_p}}{2}. \qquad (6.32)$$

Besides, in the selection of poles γ_1 and μ, the conditional equation (6.11) of velocity limitation of the servo motor and the torque limitation of the servo motor should be considered. The torque limitation of the servo motor is described as

$$C \left| K_v \left[K_p\{u(t) - y(t)\} - \frac{dy(t)}{dt} \right] \right| \leq T_{\max} \qquad (6.33)$$

where $T_{\max}$ denotes the maximum torque of the servo motor and C the coefficient of transformation from acceleration to torque. These parameters are the fixed values of the instrumentation. Through the computer simulation, the poles γ_1 and μ are satisfied (6.11), (6.32) and (6.33) with minimum are selected.

6.1.2 Properties Analysis of the Modified Taught Data Method

The introduced modified taught data method in this section is based on the theory of the pole assignment regulator. The regulator theory is always used in order to let the objective point reaching the system output. However, the control of the mechatronic servo system is the following control, i.e., the objective trajectory is time-variable. Besides, in the derivation of the modification element, the assumption $dr(t)/dt \simeq 0$ is introduced when using the 1st order model and the assumption $d^2r(t)/dt^2 + K_v dr(t)/dt \simeq 0$ is introduced when using the 2nd order model in order to adopt the pole assignment regulator theory. However, these assumptions are not often satisfied actually for the objective trajectory when considering the utilization conditions of mechatronic servo system. Therefore, the meaning of introducing these assumptions should be discussed. The improvement of the response properties of using the modified taught data method and that of using the conventional method with the original objective trajectory in the taught data should be compared in the time domain and frequency domain.

(1) The 1st Order Model

The properties analysis of the modified taught data method based on the 1st order model is discussed. Firstly, the analysis is made in the time domain. Based on the inverse Laplace transform (refer to the appendix A.1), the equation on the relationship between the objective trajectory $r(t)$ of the modified taught data method and the output $y(t)$ of the control system in the time domain can be changed from the transfer function (6.9) to

$$\frac{dy(t)}{dt} = \gamma y(t) - \gamma r(t). \tag{6.34}$$

On the other hand, based on the inverse Laplace transformation, the equation which describes the properties of the objective trajectory $r(t)$ and output $y(t)$ when the values of the objective trajectory is directly used as the taught data in the conventional method as $u(t) = r(t)$ can be changed from the transfer function (6.3) to

$$\frac{dy(t)}{dt} = -K_p y(t) + K_p r(t). \tag{6.35}$$

With the comparison between the properties of the modified taught data method (6.34) and that of the conventional method (6.35), the coefficient of $-y(t)$ and $r(t)$ can be changed from K_p to $-\gamma$. Namely, in the modified taught data method, the properties of the system are transformed from K_p to $-\gamma$ according to the proper taught data. In order to design properly the pole of the regulator γ in the scale of $\gamma < -K_p$, where the time constant of (6.34) is $-1/\gamma$, the time constant $1/K_p$ of (6.35) in the conventional method can become smaller. Therefore, the output $y(t)$ can trace the objective trajectory $r(t)$ quickly with the small time constant in the modified taught data method. If with the same precision of the contour control, the velocity of the objective trajectory in the proposed method is increased to $-\gamma/K_p$ times than that in the conventional method.

Next, the analysis is made in the frequency domain. Fig. 6.4 shows the Bode diagram under the conditions of $K_p = 15[1/\text{s}]$, $\gamma = -60[1/\text{s}]$. The Bode diagrams of the system before revision as Fig. (a) and that of the system after revision as Fig. (b) are compared. From the Bode diagram of the system after revision in Fig. (b), the frequency considered with a boundary is $\omega = 30$ [rad/s] when the gain property is constant at 0 [dB]. This frequency is higher than the $\omega = 7$ [rad/s] of the gain property of the control system of the mechatronic servo system in Fig. (a). Concerning the phase characteristics, the boundary frequency $\omega = 1$ [rad/s] at which there nearly does not generate time delay is higher comparing with $\omega = 0.02$ [rad/s] in Fig. (a). With these improvements in properties by the revision of the taught data, the cut-off frequency can be changed from $-K_p$ to $-\gamma$. The gain properties of the modification element is changed from (6.8) as

$$|F_1(j\omega)| = -\frac{\gamma}{K_p}\sqrt{\frac{\omega^2 + K_p^2}{\omega^2 + \gamma^2}}. \tag{6.36}$$

From the gain property of Fig. (c), the gain of the modification element begins to increase accompanying the increase of frequency near $\omega = 7$ [rad/s] and reaches about 12 [dB] at $\omega = 500$ [rad/s]. This frequency $\omega = 7$ [rad/s] from which the gain of the modification element begins to increase is the same as the frequency from which the gain of the mechatronic servo system begins to drop. This phenomenon of the modification element describes the compensation of the gain of the control system in the original mechatronic servo system.

6.1 Modified Taught Data Method Using a Mathematical Model 131

Besides, the phase characteristics of the modification element is changed from (6.8) to

$$\arg F_1(j\omega) = -\tan^{-1}\frac{(\gamma + K_p)\omega}{\omega^2 - K_p\gamma}. \qquad (6.37)$$

With the phase characteristics in Fig. (c), the modification element can cause the phase to advance in the high frequency band comparing with $\omega = 0.02$ [rad/s]. This frequency is identical with the frequency whose phase of the control system in the mechatronic servo system begins the delay. The maximum phase of the modification element can be calculated as

$$\sin\phi_m = \frac{\gamma + K_p}{\gamma - K_p}. \qquad (6.38)$$

The frequency at this moment is changed as [28]

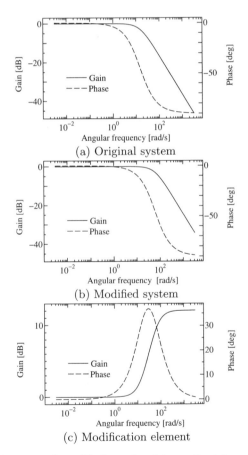

Fig. 6.4. Bode diagram of modified taught data method based on the 1st order model ($K_p = 15[1/s]$, $\gamma = -60[1/s]$)

$$\omega_m = \sqrt{-K_p\gamma}. \qquad (6.39)$$

From the above analysis, the modification element brings about the phase-lead compensation. Because of it and according to the modification element, the mechatronic servo system does not generate the gain deterioration and phase delay and also traces the objective trajectory quickly facing to the objective trajectory including the high-frequency factors compared with the conventional method using the original objective trajectory in the taught data.

Comparing with the previously adopted feedback control by inverse dynamics with the modification element $F_1(s)$ in the feedforward control by inverse dynamics, the modified taught data will be diverse when the objective trajectory cannot be differentiated. Facing this problem, the modified taught data cannot be differentiated from the proper modification element from equation (6.8) in the modified taught data method. Besides, in the limit of $\gamma \to -\infty$, the modified taught data method corresponds to the feedforward control by inverse dynamics.

In addition, comparing the revised taught data based on the servo theory without using the assumption $dr(t)/dt \simeq 0$, the proposed method based on the pole assignment regulator using the assumption $dr(t)/dt \simeq 0$ is predominance.

The differential equation about the taught data, which is represented in the 2nd order state space of systems with one integrator, constructed b the 1st order servo based on the minimum order observer (refer to the appendix A.4) and pole assignment regulator (refer to the appendix A.3) and equivalent to the equation (6.7) derived by the pole assignment regulator, can be derived as

$$\frac{d^3u(t)}{dt^3} + a_2\frac{d^2u(t)}{dt^2} + a_1\frac{du(t)}{dt} + a_0 u(t) = b_2 \frac{d^2 r(t)}{dt^2} + b_1 \frac{dr(t)}{dt} + b_0 r(t) \quad (6.40)$$

$$\begin{aligned}
a_0 &= lK_p^2(f_1 + f_2) \\
a_1 &= K_p(lK_p + f_1 + f_2 + lf_2) \\
a_2 &= lK_p + K_p + f_2 \\
b_0 &= lK_p^2(f_1 + f_2) \\
b_1 &= K_p(f_1 + lf_1 + 2lf_2) \\
b_2 &= f_1 + lf_2
\end{aligned}$$

where f_1 and f_2 are calculated by the poles of server system γ_1, γ_2 in the feedback gain as

$$f_1 = K_p + \gamma_1 + \gamma_2 + \frac{\gamma_1\gamma_2}{K_p} \qquad (6.41a)$$

$$f_2 = -K_p - \gamma_1 - \gamma_2. \qquad (6.41b)$$

l has the relationship with the pole of the observer μ in the design of the parameter as

$$\mu = -lK_p. \tag{6.42}$$

The transfer function $G_s(s)$ of the whole control system using the 1st order servo can be described by the third order system with zero as

$$G_s(s) = \frac{K_p(c_1 s + c_0)}{s^3 + a_2 s^2 + a_1 s + a_0} \tag{6.43}$$

$$c_0 = lK_p(f_1 + f_2)$$
$$c_1 = f_1 + lf_2.$$

The poles of $G_s(s)$ are γ_1, γ_2, μ and the zeros are $\gamma_1\gamma_2\mu/\{(K_p+\gamma_1+\gamma_2)(K_p+\mu)+\gamma_1\gamma_2\}$. Comparing with the zeros of $G_s(s)$ and the real parts of the poles, overshoot will be generated when the zeros are always bigger than that of the real parts of the poles.

For this case, the modified taught data method with a servo theory has the shortcoming of generating an overshoot when the following a trajectory tracing the objective trajectory comparing it with the modified taught data method with the pole assignment regulator and the properties of tracing the time variation of the objective trajectory can be found. Therefore, the modified taught data method based on the pole assignment regulator theory shows the predominance because the correct locus expressed by the arm position is very important in the contour control of the mechatronic servo system and the generation of an overshoot is the fatal shortcoming.

(2) The 2nd Order Model

In this part, the properties analysis of the modified taught data method based on the 2nd order model is made. The properties in the 2nd order model is almost that same as that based on the 1st order model. In the time domain, the modification element transformed the poles of the mechatronic servo system from $(-K_v \pm \sqrt{K_v^2 - 4K_vK_p})/2$ to γ_1 and μ comparing the original mechatronic servo system (6.12) with the mechatronic servo system after revision (6.31). In the frequency domain, the Bode diagram of the modified taught data method is based on the 2nd order model with the parameters of $K_p = 15[1/s]$, $K_v = 60[1/s]$, $\gamma_1 = \gamma_2 = -60[1/s]$, $\mu = -120[1/s]$ is shown in Fig. (6.5). It is almost the same with the properties based on the 1st order model shown in (6.4). The modified taught data method is based on the 2nd order model can be also regarded as the phase-lead compensator.

6.1.3 Experimental Verification of the Modified Taught Data Method

In order to verify the effectiveness of the modified taught data method, an experiment was made with the six-freedom-degree robot arm (Performer K10S;

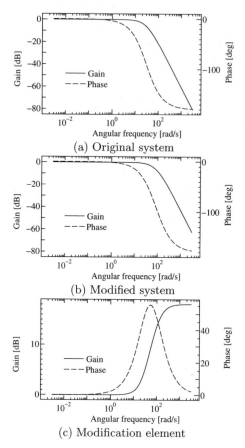

Fig. 6.5. Bode diagram of modified taught data method based on the 2nd order model ($K_p = 15[1/s]$, $K_v = 60[1/s]$, $\gamma_1 = \gamma_2 = -60[1/s]$, $\mu = -120[1/s]$)

please refer to the experiment instrumentation E.3). The position loop gain of the Performer and its velocity loop gain are $K_p = 15[1/s]$ and $K_v = 60[1/s]$, respectively. The torque limitation is $T_{\max} = 1.0[\text{Nm}]$ with a velocity limitation of the servo motor $V_{\max} = 1[\text{m/s}]$, and the coefficient of transformation from acceleration to torque is $C = 5.3 \times 10^{-3}[\text{kgm}]$. Installing the pen at the tip of the robot arm, an experiment has been made with drawing the two-dimensional trajectory at the robot arm.

The method of generation of the revised taught data is that, firstly, the revised taught data $u(t)$ was calculated with the solution of the differential equation based on the 1st order model (6.7) and the differential equation based on the 2nd order model (6.25). In the solution of the differential equation, the Euler method was used. The taught position was derived from the sampled taught data $u(t)$ with a time interval of 20[ms]. Additionally, the taught velocity was calculated by taking the discreteness of the continuous taught position.

Fig. 6.6 shows the experimental result. The objective trajectory is as the left top part of Fig. 6.6 which contains three line segments and two angles. The velocity of the objective trajectory is 250[mm/s]. Fig. 6.6 shows the experimental results with three methods. The poles of the regulator and the observer were $\gamma = -60[1/s]$ based on the 1st order model and $\gamma_1 = -60[1/s]$, $\gamma_2 = -60[1/s]$, $\mu = -120[1/s]$ based on the 2nd order model in the computer simulation.

In the following locus shown in Fig. (a) used in the conventional method, there was the movement delay of the robot arm at the angle. In the following locus using the modified taught data method based on the 1st order model as Fig. (b) or the 2nd order model as Fig. (c), the delay of the robot arm has been properly compensated and traced the angles correctly. However, the overshoot can be found in the results based on the 1st order model. In the contour control of the mechatronic servo system, this kind of overshoot should be avoided (refer to the 1.1.2 item 3). Therefore, from the results based on the 2nd order model, the overshoot has disappeared and the following locus was identical with the original objective locus. The reasons for generating an overshoot in the results based on the 1st order model, are that the modeling error cannot be neglected when the robot arm was modeled by the 1st order model with the objective velocity 250[mm/s].

Comparing the surface area of the errors between the objective locus and the following locus, in the conventional method is 136[mm^2], in the modified taught data method based on the 1st order model is 60[mm^2], and in the modified taught data method based on the 2nd order model is 40[mm^2]. From these results, the effectiveness of the modified taught data method was verified.

6.2 Modified Taught Data Method Using a Gaussian Network

In the modified taught data method based on the model in the previous section, the servo parameters K_p, K_v in the model are necessary to be correctly identified in advance.

In the modified taught data method based on one type of neural network, the Gaussian network, and the information of the movement with the test pattern, the identification of the mechatronic servo system can be realized by the Gaussian network as equation (6.46). The revision by taught data based on this kind of Gaussian network can be also conducted.

Although the role of the taught data revision is the same as the method based on the model in the former section, the merit of this method based on the Gaussian network is that the characteristics of the mechatronic servo system need not be known in advance.

136   6 The Modified Taught Data Method

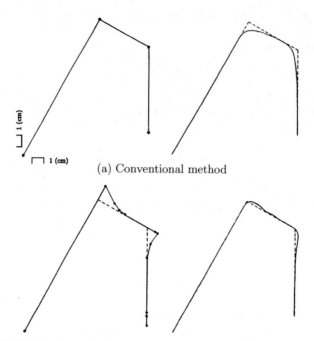

(a) Conventional method

(b) Modified taught data method based on the 1st order model

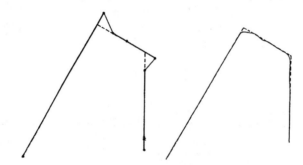

(c) Modified taught data method based on the 2nd order model

Fig. 6.6. Experimental results by using industrial robot. The left figures are about taught data and the right figures are following locus.

6.2.1 Derivation of Modified Taught Data Method Using a Gaussian Network

(1) Principle of the Modified Taught Data Method

The modified taught data method is to compensate for the delay of the mechatronic servo system by the taught data which is the input of the servo system (refer to the 6.1.1). When revising the taught data and the modeling of the servo system is correct, although the modification element can be constructed for revising the taught data based on the above model and it is possible to obtain the high-precision contour control, it is difficult to obtain the correct general model and there are always many modeling errors in the equation (6.45). Therefore, with neural networks and learning from the inverse system, control performance can be improved. In this section, through using a Gaussian network, the modification element can be constructed by learning the actual dynamics of servo system.

(2) A Mathematical Model of the Mechatronic Servo System

The mathematical model of a mechatronic servo system which is necessary for the construction of the Gaussian network and determination of the initial parameters will be introduced. As the mathematical model of mathematic servo system, the 2nd order model which approximates the actual servo system until the velocity loop is adopted (refer to 2.2.4). The equation of the 2nd order model is shown as

$$\frac{d^2y(t)}{dt^2} = -K_v\frac{dy(t)}{dt} - K_pK_vy(t) + K_pK_vu(t) \tag{6.44}$$

where $u(t)$ denotes the position input to the servo system, $y(t)$ denotes the position output to the servo system and K_p, K_v have the meaning of K_{p2}, K_{v2} of the middle speed 2nd order model as in equation (2.29) in section 2.2.4, respectively. Also, the construction of the inverse dynamics of the servo system by the Gaussian network is based on the inverse solution of equation (6.44) with $y(t) = r(t)$, which $r(t)$ denotes the objective trajectory

$$u(t) = r(t) + \frac{1}{K_p}\frac{dr(t)}{dt} + \frac{1}{K_vK_p}\frac{d^2r(t)}{dt^2}. \tag{6.45}$$

This mathematical model expresses the characteristics of the servo system. However, the real parameters have the difference with the setting values for products. Also, the nonlinear terms which cannot be expressed by the 2nd order model exist in the dynamics. Therefore, the modeling error is assumed to exist in the inverse dynamics of equation (6.45), and the learning from the inverse dynamics of the servo system by the Gaussian network will conduct.

(3) Construction of Inverse Dynamics by the Gaussian Network

(i) Gaussian network

The Gaussian network is a type of neural network whose units use a Gaussian function (Gaussian unit)[30]. As the characteristics of the Gaussian units, the output of the units is toward the input around the mean. If the input leaves the mean, the output of the unit approaches to 0. Though Gaussian units which are the components of a general Gaussian network possess multiple inputs, in order to simplify the structure, the part about the mutual correlation of the Gaussian function in the Gaussian unit which is as one input, are all regarded as zero, and one input and one output Gaussian unit is used.

In this section, the adopted Gaussian network is composed of multiple units. From the following equation

$$\phi(\boldsymbol{x}) = \sum_{i=1}^{M} w_i \psi_i(x_i) \tag{6.46}$$

each unit is the one input Gaussian unit

$$\psi_i(x_i) = \exp\left\{-\frac{(x_i - m_i)^2}{2\sigma_i^2}\right\} \tag{6.47}$$

where $\boldsymbol{x} = (x_1, \cdots, x_M)$ is the input of the network, $\phi(\boldsymbol{x})$ is the output of the network, M is the number of units, w_i is the weight of the ith unit, $\psi_i(x_i)$ is the output of the ith unit, m_i is the mean of the ith unit, and σ_i is the standard deviation of the ith unit. According to the equation of the Gaussian network (6.46), the inverse dynamics of the actual servo system can be constructed.

(ii) Determination of the structure

As shown in the Fig. 6.7, the adopted Gaussian network has three layers, three input, six intermediate units and one output. In the structure shown in Fig. 6.7, three inputs of the Gaussian network are realized by the six intermediate units $\boldsymbol{x} = (x_1, \cdots, x_6)$ in which every two units have the same input. The inputs of the network $(r, dr/dt, d^2r/dt^2)$, in another word, are the $x_1 = x_2 = r$, $x_3 = x_4 = dr/dt$, $x_5 = x_6 = d^2r/dt^2$. The first item of the right-hand side of the inverse dynamics equation (6.45) is approximated by the first and second units, the second item by the third and fourth units and the third item by the fifth and sixth units. The output of the Gaussian network is regarded as the input of the servo system for the revised taught data. In the Fig. 6.7, • denotes the Gaussian unit and ○ denotes the linear unit.

(iii) Determination of the initial parameter

In order to approximate the inverse dynamics of (6.45) by a Gaussian neural with the initial parameters, the initial parameters should be determined.

6.2 Modified Taught Data Method Using a Gaussian Network

In the determination of the initial parameters, the Gaussian network shown in the Fig. 6.7 should be divided into three parts and the one-input, two-intermediate-unit and one-output Gaussian network is considered. In this Gaussian network, the symbols of the means of the two units are changed as below in order to approximate the general linear function $y = ax$,

$$\phi(x) = w\exp\left\{-\frac{(x-m)^2}{2\sigma^2}\right\} - w\exp\left\{-\frac{(x+m)^2}{2\sigma^2}\right\}. \tag{6.48}$$

Equation (6.48) is approximated by the one-order Taylor expansion,

$$\phi(x) \approx \frac{2wm}{\sigma^2}\exp\left(-\frac{m^2}{2\sigma^2}\right)x. \tag{6.49}$$

If the inclination of the linear function is as

$$a = \frac{2wm}{\sigma^2}\exp\left(-\frac{m^2}{2\sigma^2}\right) \tag{6.50}$$

the linearization in the neighborhood of $x = 0$ can be realized. The variation of the relationship between the standard deviation σ and the mean m can be described in Fig. 6.8. With the results when the coefficient which hung on mean m is changed in each 0.01 and $\sigma = 0.57m$, $\phi(x)$ can approximate the ax in the scale of x. In another words, the linear function $y = ax$ can be approximated when the parameters of the Gaussian network are as below and the $\phi(x)$ of Gaussian network is linearized within the $x_{\max}$, and equation (6.50) and $\sigma = 0.57m$ are used,

$$m = x_{\max}, \quad \sigma = 0.57x_{\max}, \quad w = 0.757ax_{\max}. \tag{6.51}$$

At this moment, the minimum of $\phi(x)$ is $-0.755ax_{\max}$ with $x = -x_{\max}$ and the maximum is $0.755ax_{\max}$ with $x = x_{\max}$.

Using this relationship, the initial parameters of the whole three-inputs, six-units and one-output Gaussian network can be give as

$$\left.\begin{array}{l} m_1 = -m_2 = x^p_{\max} \\ \sigma_1 = \sigma_2 = 0.57x^p_{\max} \\ w_1 = w_2 = 0.757x^p_{\max} \end{array}\right\} \tag{6.52a}$$

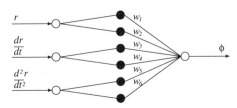

Fig. 6.7. Structure of Gaussian network

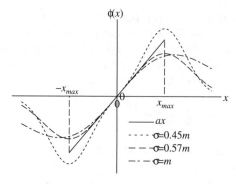

Fig. 6.8. Determination of initial parameters of Gaussian network

$$\left.\begin{array}{l} m_3 = -m_4 = x^v_{\max} \\ \sigma_3 = \sigma_4 = 0.57 x^v_{\max} \\ w_3 = w_4 = \dfrac{0.757 x^v_{\max}}{K_p} \end{array}\right\} \quad (6.52b)$$

$$\left.\begin{array}{l} m_5 = -m_6 = x^a_{\max} \\ \sigma_5 = \sigma_6 = 0.57 x^a_{\max} \\ w_5 = w_6 = \dfrac{0.757 x^a_{\max}}{K_p K_v} \end{array}\right\} \quad (6.52c)$$

the inverse dynamics in equation (6.45) can be appropriately approximated by the initial parameters of the Gaussian network. where $x^p_{\max}$, $x^v_{\max}$, $x^a_{\max}$ in the equation (6.52) denote the detail variables of the parameters $x_{\max}$ designed by linearizable regions for the position, velocity and acceleration, respectively. Like this, even the input signal of the Gaussian network has a big error over the input scale of the servo system, the safety of the instrument can be guaranteed because the output of the Gaussain network can be changed into 0 owing to using a Gaussian network when designing the modification elements.

(iv) Learning algorithm

Through the Gaussian network with the initial parameters, the inverse dynamics expressed in equation (6.45) can be approximated. However, since the modeling error exists in the mathematical model of the servo system expressed with the general equation (6.44), there exist errors in the inverse dynamics given by equation (6.45). In order to reduce the errors of this modeling error, the learning of Gaussian network should be preformed using the teaching signal from the experimental results when the servo system was actually moved. The loss function during the learning of Gaussian network is as

$$E_{\mathrm{rms}} = \sqrt{\dfrac{2}{L} \sum_{l=1}^{L} E^l} \quad (6.53)$$

6.2 Modified Taught Data Method Using a Gaussian Network

$$E^l = \frac{1}{2}\{u^l - \phi(\boldsymbol{x}^l)\}^2 \tag{6.54}$$

where $(u^l, \boldsymbol{x}^l) = (u^l, x_1^l, \cdots, x_6^l)$ denotes the teaching signal during the learning of the Gaussian network and L denotes the number of the teaching signal.

In the learning of Gaussian network parameters, error back propagation learning is used [31]. The variation of parameters $\boldsymbol{p}_i = (w_i, m_i, \sigma_i)$, $\Delta \boldsymbol{p}_i = (\Delta w_i, \Delta m_i, \Delta \sigma_i)$, $i = 1, \cdots, 6$ with a learning rate η can be described as

$$\boldsymbol{p}_i^{\text{new}} = \boldsymbol{p}_i^{\text{old}} + \eta \Delta \boldsymbol{p}_i, \quad i = 1, \cdots, 6 \tag{6.55}$$

$$\Delta w_i = -\frac{\partial E^l}{\partial w_i}$$
$$= \{u^l - \phi(\boldsymbol{x}^l)\}\psi_i(x_i^l)$$

$$\Delta m_i = -\frac{\partial E^l}{\partial m_i}$$
$$= \frac{(x_i^l - m_i)}{(\sigma_i)^2}\psi_i(x_i^l)\{u^l - \phi(\boldsymbol{x}^l)\}w_i$$

$$\Delta \sigma_i = -\frac{\partial E^l}{\partial \sigma_i}$$
$$= \frac{(x_i^l - m_i)^2}{(\sigma_i)^3}\psi_i(x_i^l)\{u^l - \phi(\boldsymbol{x}^l)\}w_i.$$

The learning process will stop when the loss function of equation (6.53) is below the threshold. The learning of the Gaussian network can be expressed by the functions in the structure of Fig. 6.7 for the whole parameters. With the learning, the Gaussian network can learn from the inverse dynamics of the real servo system. For example, according to the symbol of the input of the servo system, the characteristics of the servo system will be changed. Moreover, when the inclination a of linear function is changed according to the positive or negative input, the nonlinear part which cannot be expressed by the linear neural network can be realized.

(4) Utilization of the Gaussian Network

After learning, the Gaussian network is used for the modification elements. The Gaussian network cannot only express the inverse dynamics of the servo system and provide the revised taught data by its output for the servo system, but also the servo system moved by this taught data can expect that the following trajectory approaches the objective trajectory. In the learning of the inverse dynamics of the servo system by the Gaussian network, there is no need to let the objective trajectory for producing the teaching signal is same the objective trajectory in the actual operation. That is to say, after one time learning of the Gaussian network for the inverse dynamics of the servo system, the following trajectory can approach any objective trajectory when using this Gaussian network for the modification elements.

6.2.2 Experimental Verification for Modified Taught Data Method Using a Gaussian Network

(1) Conditions of the Experiment

In order to verify the effectiveness of a Gaussian network based on the 2nd order model shown in 6.2.1, the experiment of contour control using an XY table was made (refer to the experiment instrument E.4). The control of the XY table is constructed by two Gaussian networks in equation (6.46) for independent axes in order to conduct the independent movement of the x axis and the y axis, respectively. The experimental results will be shown when the objective trajectory of the XY table is as

$$u_x(t) = \begin{cases} 4.8 & (0 \leq t < 0.5) \\ 4\cos\left\{\dfrac{\pi(t-0.5)}{2}\right\} + \dfrac{4}{5}\cos\left\{\dfrac{5\pi(t-0.5)}{2}\right\} & (0.5 \leq t < 4.5) \\ 4.8 & (4.5 \leq t \leq 5) \end{cases}$$

$$u_y(t) = \begin{cases} 0 & (0 \leq t < 0.5) \\ 4\sin\left\{\dfrac{\pi(t-0.5)}{2}\right\} + \dfrac{4}{5}\sin\left\{\dfrac{5\pi(t-0.5)}{2}\right\} & (0.5 \leq t < 4.5) \\ 0 & (4.5 \leq t \leq 5). \end{cases}$$

(2) Generation of the Teaching Signal

In the determination of the initial parameters of the Gaussian network, the defined value $K_p = 5[1/s]$ of the position loop gain of the equipment in the equation (6.52) was used, and the critical condition from $K_v = 4K_p$ to $K_v = 20[1/s]$ of the velocity loop gain used in the industrial field, which cannot be defined directly, was used. This K_v which was not the value measured by the actual device was considered to contain large errors. But the high-precision contour control can be realized because in the proposed method the Gaussian network for the modification element was used and the inverse dynamics can be constructed based on the learning from the actual equipment.

Besides, the linearizable region condition of the equipment was considered as 15[cm] in the movable region of the table. The output scale of the two Gaussian units about the position were set as $-7.5 \leq \phi(r) \leq 7.5$[cm] when $x^p_{max} = 10$[cm]. The maximal velocity of the equipment was considered as 9.3[cm/s]. The output scale of the two Gaussian units about velocity were set as $-11.325 \leq \phi(dr/dt) \leq 11.325$[cm/s] when $x^v_{max} = 15$[cm/s]. Concerning the safety of the equipment, the output scale of the two Gaussian units about acceleration were set as $-60.4 \leq \phi(d^2r/dt^2) \leq 60.4$[cm/s^2] which was not over the maximal acceleration of 84.7[cm/s^2]. The teaching signal of learning for the above Gaussian network with initial parameters came from the output

6.2 Modified Taught Data Method Using a Gaussian Network   143

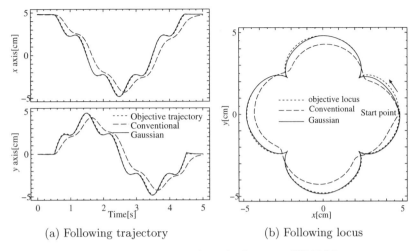

Fig. 6.9. Experimental results by using XY Table

data obtained by the computer when the properties of the servo system expressed in the movement can be given arbitrary and the XY table was moved with the original objective trajectory in the experiment. The sampling time interval is $\Delta t_p = 10[\text{ms}]$ when making the teaching signal was the same as that of the contour control experiment. Therefore, the teaching signals were obtained as $(u^l, x^l) = (u(l\Delta t_p), y(l\Delta t_p), y(l\Delta t_p), dy(l\Delta t_p)/dt, dy(l\Delta t_p)/dt, d^2y(l\Delta t_p)/dt^2, d^2y(l\Delta t_p)/dt^2)$, $l = 0, \cdots, 500$. However, the data obtained by the computer from the actual XY table were only the velocity output dy/dt of the techogenerator obtained from the servo motor. The position output y was the numerical integral of the velocity output and the acceleration output $\ddot{y}$ was the numerical differential of the velocity output. Additionally, the velocity output dy/dt of the techogenerator were the results whose noise have been deleted by the band pass filter of $0 \sim 10[\text{Hz}]$. With the learning rate of $\eta = 0.001$ during the Gaussian network learning, the learning process will stop when the common threshold of the x axis and the y axis was below the $0.35[\text{mm}]$. There were 182 learning times when the data set of the teaching signal (u^l, x^l), $l = 0, \cdots, 500$ was regarded as one time learning.

(3) Experimental Results of the Contour Control

By using the Gaussian network shown in the Fig. 6.7 after learning, the experimental results of contour control with the input of the XY table using the revised taught data revised by the Gaussian network were shown. Fig. 6.9(a) shows the following trajectory of the experimental results in the Gaussian network after learning. Fig. 6.9(b) shows the following locus in the XY plate. Here, the objective trajectory without any revision was used in the conventional method. Comparing with the conventional method without any revi-

144   6 The Modified Taught Data Method

sion, the following trajectory was regarded as the following locus was clearly approaching the objective when using the Gaussian network to realize the revision. Therefore, the high-precision control can be realized.

6.3 A Modified Taught Data Method for a Flexible Mechanism

When the movement of the robot arm becomes faster, the flexible mechanism of the robot arm is necessary for the flexibility of the manipulator and flexible connection of the link. If neglecting the characteristics of flexibility, oscillation or overshoot in the movement of the robot arm will occur. The contour control performance will deteriorate and the determination time of the position will increase.

According to the flexible mechanism, the mathematical model is made. Based on this equation, the taught data modification element of the former section is constructed. The high-precision contour control can be realized in the robot manipulator of the flexible mechanism.

Then, the requirement of a high-speed, high-precision movement of a manipulator in industry, the proposed technique as the control method which can bring the current system into maximal effect is very important without huge change of hardware in the current system.

6.3.1 Derivation of Contour Control with Oscillation Restraint Using the Modified Taught Data Method

In order to realize contour control with oscillation restraint in the movement of the flexible arm, the block diagram of the control system in the one axis flexible arm shown in 6.10 is considered. In the Fig. 6.10, $R(s)$ denotes the objective trajectory, $Z(s)$ denotes the position of the arm fulcrum, $Y(s)$ denotes the output (tip position of the arm), K_p denotes the position loop gain. The modified taught data method (refer to 6.1.1) is adopted with the modification element $F_3(s)$ for constructing the taught data revised from the objective trajectory of arm. In this section, although only one axis is considered, the realization of control with oscillation restraint for one axis can also be adapted for the multi-axis mechatronic servo system.

The dynamics of the servo system which causes the movement of the arm is expressed by the 1st order model (refer to the 2.2.3). The flexible arm of the elasticity body is expressed by the 2nd order system, where ζ_L denotes the damping factor and ω_L denotes the natural angular frequency. Therefore, the whole transfer function of the control system of this flexible arm is expressed as

6.3 A Modified Taught Data Method for a Flexible Mechanism

$$G_3(s) = \frac{a_0}{s^3 + a_2 s^2 + a_1 s + a_0} \tag{6.56}$$

$$a_0 = K_p \omega_L^2$$
$$a_1 = \omega_L^2 + 2\zeta_L \omega_L K_p$$
$$a_2 = K_p + 2\zeta_L \omega_L.$$

In the modified taught data method, the modification element $F_3(s)$ is derived using the pole assignment regulator and the minimum order observer for the control system to solve the characteristics of the closed-loop system and transfer it to the open-loop system whose relationship of the input and output is equivalent to the transfer function of the closed-loop system. For the control system of equation (6.57), the modification element is as

$$F_3(s) = \frac{b_5 s^5 + b_4 s^4 + b_3 s^3 + b_2 s^2 + b_1 s + b_0}{(s-\gamma_1)(s-\gamma_2)(s-\gamma_3)(s-\mu_1)(s-\mu_2)} \tag{6.57}$$

$$b_0 = a_0(h_0 - g_0)$$
$$b_1 = a_0(h_1 - g_1) + a_1(h_0 - g_0)$$
$$b_2 = a_0(1 - g_2) + a_1(h_1 - g_1) + a_0(h_0 - g_0)$$
$$b_3 = a_1(1 - g_2) + a_2(h_1 - g_1) + h_0 - g_0$$
$$b_4 = a_2(1 - g_2) + h_1 - g_1$$
$$b_5 = 1 - g_2$$
$$g_0 = l_2 f_1 + (l_1 l_2 + k_2) f_2 + (l_2^2 + l_1 k_2 - l_2 k_1) f_3$$
$$g_1 = l_1 f_1 + (l_1^2 + k_1) f_2 + (l_1 l_2 + k_2) f_3$$
$$g_2 = f_1 + l_1 f_2 + l_2 f_3$$
$$h_0 = l_2 - a_0 f_2 - a_0 l_1 f_3$$
$$h_1 = l_1 - a_0 f_3$$
$$l_1 = -(\mu_1 + \mu_2)$$
$$l_2 = \mu_1 \mu_2$$
$$k_1 = -l_1^2 + l_2 - a_1 + a_2 l_1$$
$$k_2 = -l_1 l_2 - a_0 + a_2 l_2$$

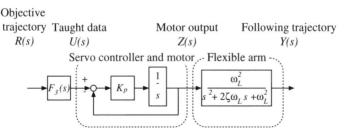

Fig. 6.10. Block diagram of modified taught data method for flexible arm

$$f_1 = -(d_1 - a_2 d_2 + (a_2^2 - a_1)d_3 - a_0 - a_2^3 + 2a_1 a_2)/a_0$$
$$f_2 = -(d_2 - a_2 d_3 - a_1 + a_2^2)/a_0$$
$$f_3 = -(d_3 - a_2)/a_0$$
$$d_1 = -\gamma_1 \gamma_2 \gamma_3$$
$$d_2 = \gamma_1 \gamma_2 + \gamma_2 \gamma_3 + \gamma_3 \gamma_1$$
$$d_3 = -(\gamma_1 + \gamma_2 + \gamma_3).$$

In the equation (6.57), the modification element expressed by the 1st order transfer function for the rigid body system shown in 6.1.1 is expanded into the fifth-order modification element including the observer. γ_1, γ_2, γ_3 are the poles of the regulator and μ_1, μ_2 are the poles of the minimal order observer. From the taught data $u(t)$ generated through the modification element $F_3(s)$, tracing correctly the objective trajectory without oscillation in the flexible arm can be realized.

6.3.2 Experimental Verification of Oscillation Restraint Control Using the Modified Taught Data Method

Through the experimental device of the flexible arm which emphasizes the arm elasticity characteristic of one axis of the mechatronic servo system, the effectiveness of the proposed method can be verified. With the metal plate in the flexible arm, the bottom edge of this flexible arm is installed in the base seat of the drive device which consists of combinations with a DC servo motor and the ball screw. The control purpose is to make the flexible arm correspond to the objective trajectory without the oscillation from the static state of the base seat to another static state after moving to the objective position. The size of the metal board is as follows, the length is 0.83[m], width is 0.028[m] and height is 0.002[m]. The mass is 351[g], the elasticity coefficient is $K = 73785.2[g/s^2]$, the viscous frictional coefficient is $D_L = 3.626[g/s]$, the natural angular frequency is $\omega_L = 14.5[Hz]$, the damping factor is $\zeta_L = 3.56 \times 10^{-4}$, and the position loop gain is $K_p = 15[1/s]$. The objective trajectory is the moving trajectory with the velocity of 0.03[m/s]. The design parameters in the equation (6.57) are the poles of the regulator $\gamma = -10$ (three-fold root) and the poles of the observer $\gamma = -20$ (two-fold root).

Fig. 6.11 shows the experimental results of the proposed method with the equivalent velocity movement with 0.03[m/s] of the base seat. The horizontal axis of the graph is time and the vertical axis is the oscillation in the center of gravity of flexible arm. From the results of the oscillation in the Fig. (a) with the modified taught data method of the proposed method, the maximal amplitude is 0.45[mm]. The maximal value of the oscillation in the results of the equivalent velocity movement in Fig. (b) is 2.0[mm]. Comparing with one another, the amplitude of oscillation in the center of gravity of the arm is reduced to the 1/4. The left oscillation is from the modeling error which cannot be generated in the ideal simulation results.

6.3 A Modified Taught Data Method for a Flexible Mechanism

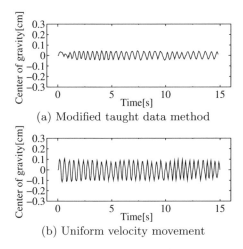

Fig. 6.11. Experimental result

The adaptiveness possibility of the modeling error of the modified taught data method was investigated. With the simulation, the scale of the oscillation arm when the design error is put in the damping factor ζ_L or the natural angular frequency ω_L was calculated. When the size of the oscillation of the arm with the put design error was within the allowance of modeling error in order to let it below 10[%] of the maximal oscillation without design error, and the natural angular frequency ω_L is $-4.1 \sim 2.8$[%], then the size of the oscillation became $-100 \sim 3549$[%] in the damping factor ζ_L.

7
Master-Slave Synchronous Positioning Control

When one robot manipulator has many links and each of them corresponds to one axis of the motor, it is very important to realize the synchronous positioning of each axis in the high-precision contour control. In this chapter, we propose a new high-precision contour control not subject to the restriction of the current conditions. It is adapted for the master-slave synchronous positioning control, which supposes one axis as the master-axis and another as the slave-axis without a large characteristic value K_p of the servo system.

7.1 The Master-Slave Synchronous Positioning Control Method

The typical applications which requires synchronous movement based on the relationship between the master axis and the slave axis are tapping process work, installing tapping tools in the rotated master axis and processing screw by an up and down movement of master axis (sending) with rotation, and so on. Since the process specification of the screw pitch of the product is regular, if the rotations of the master axis and sending position are not synchronous, the screw pitch will be changed, or tools will be broken an the extreme case.

The master-slave synchronous positioning method is to generate modification term of inverse dynamics for the servo system and with this modification term, the position output of the master axis is taken as the input signal of the slave axis. If there mixed with disturbance in the master axis, from the proposed method, the slave-axis synchronous positioning method can be implemented properly.

The command of the servo system of each industrial robot axis is independently given. The command of the slave axis is revised by software. Therefore, since it is expected that the existing hardware is not changed and the desirable synchronous positioning can be realized, the value of any industrial application of this method is very high.

7.1.1 Necessity of Master-Slave Synchronous Positioning Control

(1) Mathematical Model of the Objective of the Master-Slave Synchronous Positioning Control

Concerning the control objective with the requirement of position synchronization, the overall control system with the control equipment and the servo system are almost all controlling master axes and slave axes independently. For the actuator, many servo motors have been used. In order to use high-performance device in the servo motors and their control equipments, the property of velocity control of the servo motor is considered as a fixed constant when the processing speed is not very high and the property of the position control is only considered (refer to 2.2.3). Therefore, the transfer function of the servo system is expressed as

$$P_x(s) = \frac{K_{px}}{s(s+K_{px})}U_x(s) + \frac{1}{s+K_{px}}D_x(s) \qquad (7.1a)$$

$$P_y(s) = \frac{K_{py}}{s(s+K_{py})}U_y(s) \qquad (7.1b)$$

where, the x axis is the master axis, the y axis is the slave axis, $P_x(s)$, $P_y(s)$ are the positions of the x axis and the y axis, $U_x(s)$, $U_y(s)$ are the velocity input reference of the x axis and the y axis, K_{px}, K_{py} have the meanings of K_{p1} in the equation (2.20) for the 1st order model written in the item 2.2.3 about the x axis and the y axis. The disturbance, expressed as $D_x(s)$, is only added in the master axis, supposed in the tap processing. The first item of equation (7.1a) describes the relationship between the velocity input $U_x(s)$ and the position output of the x axis. The second item describes the relationship between the disturbance $D_x(s)$ inputing into the x axis and position output of the x axis. The property of control system is described by K_{px}, K_{py}. Their values are determined by the structure of the hardware. In addition, $1/s$ before the servo system denotes the integral from the velocity input to the position input. The control purpose of the master-slave synchronous positioning control is to make the position output of the x axis and the y axis are synchronous, that is, to make the following equation successfully

$$P_y(s) = k_c P_x(s) \qquad (7.2)$$

where k_c is the proportional constant. If the position output of the x axis and the y axis satisfies equation (7.2), the position synchronization can be realized.

(2) Issues without Expectation of Position Synchronization

If the dynamics of the x axis and the y axis are not considered and the velocity input $U_y(s)$ of y axis is k_c times of velocity input of the x axis, the position output of the y axis is as

7.1 The Master-Slave Synchronous Positioning Control Method

$$P_y(s) = \frac{k_c K_{py}}{s(s + K_{py})} U_x(s). \tag{7.3}$$

The position output error of the y axis to the x axis, from equation (7.1a) and (7.3), is as

$$k_c P_x(s) - P_y(s) = \frac{k_c(K_{px} - K_{py})}{(s + K_{px})(s + K_{py})} U_x(s) + \frac{k_c}{s + K_{px}} D_x(s). \tag{7.4}$$

From equation (7.4), if there is no position synchronization, the position output of the x axis and the position output of the y axis are not synchronous because the position output error is not 0. Since the position loop gains of the x axis and the y axis are difference, there exists a deviation of position output. From this case, if we use velocity input reference of the x axis without change, the synchronous action cannot be realized because the position loop gains of the x axis and the y axis are not the same. In addition, without setting the compensation of the y axis for the disturbance $D_x(s)$ of the x axis is another reason for synchronization.

7.1.2 Derivation and Property Analysis of the Master-Slave Synchronous Positioning Control Method

(1) Derivation of the Master-Slave Synchronous Positioning Control Method

In the former part, the problem that the k_c times of velocity input reference of the x axis is simply used as the velocity input reference of the y axis was introduced. In order to make the position of the y axis synchronization with the position of axis x, the velocity input reference of axis x is revised for compensating the different dynamics between axis x and axis y. If the velocity input reference of axis y is performed like this, the position synchronization can be realized. However, if performing a revision in the velocity input reference of axis x is only for the velocity input reference of axis y, the compensation for disturbance in axis x cannot be implemented and the high-precision position synchronization cannot be realized. But if the position output of axis x is feedback as the position input of y, the impact of a disturbance in the axis x can be overcome by the feedback of the position output of axis x. If the only feedback in the position output of axis x without any change, the synchronization of axis x with the movement delay caused by the dynamics of axis y cannot be realized. Therefore, by using the inverse dynamics of axis y and revising the feedback signal of the position output of axis x, the position synchronization can be realized. Namely, in order to change the dynamics of axis y into 1, feedforward compensation is performed according to the inverse dynamics of axis y.

In order to realize the above properties, the inverse dynamics of the 1st order system of axis y $F_s(s)$ can be constructed as

152   7 Master-Slave Synchronous Positioning Control

$$F_s(s) = \frac{s + K_{py}}{K_{py}}. \tag{7.5}$$

The master-slave synchronous positioning control method, with the position output of axis x as the position input of axis y, can be given according to $F_s(s)$, is shown. This master-slave synchronous positioning control method is based on the prerequisite of different dynamics between axis x and axis y. It can be also used for compensation for any fatal effects of disturbance $D_x(s)$ mixed into axis x. When feedback the position output of axis x, it is assumed that there are no observational noises (In the mechatronic servo system, there are no observational noise because of the position test by pulse measurement in the encoder). Moreover, discussion is carried out with the assumption of correctly modeling the dynamics of axis y in the following part. When a modeling error exists, it is necessary to adjust correctly the value of K_{py} in equation (7.5) to minimize the modeling error. The block diagram of the master-slave synchronous positioning control method is illustrated in Fig. 7.1.

(2) Property Analysis of the Master-Slave Synchronous Positioning Control Method

The position output of axis y in the master-slave synchronous positioning control method is as

$$P_y(s) = \frac{k_c K_{px}}{s(s + K_{px})} U_x(s) + \frac{k_c}{s + K_{px}} D_x(s). \tag{7.6}$$

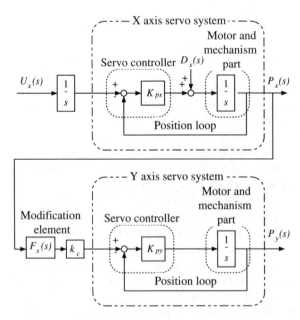

Fig. 7.1. Block diagram of master-slave synchronous positioning control method

7.1 The Master-Slave Synchronous Positioning Control Method

Comparing equation (7.1a) and (7.6), the relationship of the position output between axis x and axis y is as

$$k_c P_x(s) - P_y(s) = 0. \tag{7.7}$$

It satisfies the condition of equation (7.2). Namely, axis y is synchronized on position with axis x although the disturbance is input into axis x. However, it is necessary to make the initial value synchronization in order to coordinate with the time response for equation (7.7) in the frequency domain.

From the discussion of the realization of this method, it is necessary to confirm that the input of axis y after revision does not diverge when the modification element $F_s(s)$ contains a differential. Therefore, the position input signal of axis y should be calculated as

$$F_s(s)P_x(s) = \frac{K_{px}(s + K_{py})}{K_{py}s(s + K_{px})} U_x(s) + \frac{s + K_{py}}{K_{py}(s + K_{px})} D_x(s). \tag{7.8}$$

In order to possess the common proper transfer function (the times of denominator polynomial is bigger than that of molecule polynomial), the transfer function of the position input $F_s(s)P_x(s)$ of axis y in connection with the velocity input reference $U_x(s)$ of axis x and disturbance $D_x(s)$ in axis x is for avoiding the divergence of the position input reference of axis y. Therefore, there is no problem when using (7.6) as the modification element $F_s(s)$ and the effectiveness of the master-slave synchronous positioning control method can be verified.

7.1.3 Experimental Test of the Master-Slave Synchronous Positioning Control Method

By using the master-slave synchronous positioning control method, the effectiveness of the position synchronization of axis x and axis y can be verified using computer simulation and an experiment by using XY table (refer to E.4 about experimental equipment). The conditions of the simulation and the experiment are a position loop gain of axis x $K_{px} = 5[1/s]$, position loop gain of axis y $K_{py} = 15[1/s]$, proportional constant $k_c = 1$ and sampling time interval $\Delta t_p = 0.02[s]$.

(1) Simulation of the Master-Slave Synchronous Positioning Control

There are two kinds of supposed disturbances in the required equipment when performing position synchronization. Concerning these disturbances, simulation is made with (a) master-slave synchronous positioning control method, (b) without expectation of position synchronization and (c) a tracking control method between two servo systems [35]. The tracking control method between two servo systems in (c) is the method used to compensate for the velocity input of axis x by the position output feedback of axis x. The velocity input contains the features of the ramp and the step. After cutting the screw and returning to a trapezoidal wave as in Fig. 7.2, it is function as

$$u_x(t) = \begin{cases} 90t & (0 \leq t \leq 0.6) \\ 54 & (0.6 < t \leq 1.2) \\ -90t + 162 & (1.2 < t \leq 1.8) \\ 0 & (1.8 < t \leq 2.0, 3.8 < t \leq 4.0) \\ -90t + 180 & (2.0 < t \leq 2.6) \\ -54 & (2.6 < t \leq 3.2) \\ 90t - 342 & (3.2 < t \leq 3.8). \end{cases}$$

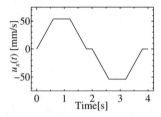

Fig. 7.2. Input trajectory (trapezoidal wave)

(i) *Step disturbance*

The step disturbance is generated when using the force with a step shape at the moment of cutting the screw in the tab processing. Based on the simulation, the step disturbance is as

$$d_x(t) = \begin{cases} 0 & (0 \leq t \leq 0.5, 2.0 < t \leq 4.0) \\ -5 & (0.5 < t \leq 2.0). \end{cases}$$

Its wave is shown in Fig. 7.3.

In order to compare the master-slave synchronous positioning control method with step disturbance, the simulation results of the tracking control method between the servo system without position synchronization is shown in Fig. 7.4. From the left side, the locus of the XY table, time change of axis x and y and trajectory error $e(t) = p_x(t) - p_y(t)$ of axis x and axis y are illustrated.

Fig. 7.4(b) illustrates the results without position synchronization for increasing the response of axis y compared with that of axis x. In this case, the maximal trajectory error is 8[mm] among the different large position loop gains of axis x and axis y as well as a different response velocity. In addition, for the big errors with different position loop gains, it cannot be seen that

7.1 The Master-Slave Synchronous Positioning Control Method   155

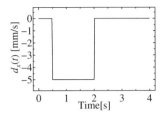

Fig. 7.3. Step disturbance

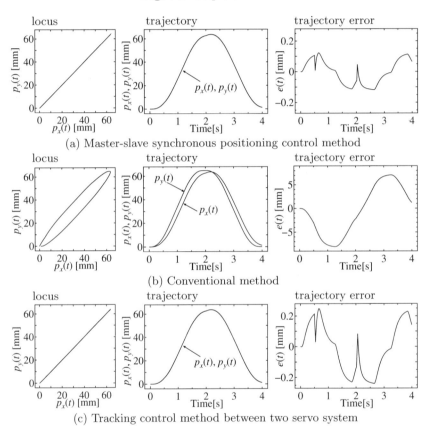

Fig. 7.4. Simulation results on step disturbance

the impact of step disturbance input between 0∼2[s] from the graph of the trajectory error (amplified with 25 times).

Comparing Fig. 7.4(a) and (c), two methods are making position synchronization as long as looking the graph of locus and time change of the XY table. Additionally, in the two methods, the impact of any step disturbance input between 0∼2[s] shown in the trajectory errors is quite small at 0.1[mm] in Fig. (a) compared with 0.25[mm] in Fig. (c). Moreover, the locus error

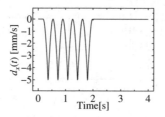

Fig. 7.5. Disturbance wave with Saw tooth state cycle

out of the moment of mixing the step disturbance in Fig. (c) is bigger than that of Fig. (a). Besides, in the situation without step disturbance between 2~4[s], the trajectory error of 0.25[mm] in Fig. (c) is bigger than the 0.1[mm] in Fig. (a). From the above comparisons, the effectiveness of the master-slave synchronous positioning control method is verified. The impact of a disturbance in master-slave synchronous positioning control method is due to the different operation in the computer for the controller for the differential of inverse dynamics $F_s(s)$ expressed in equation (7.6).

(ii) Saw-tooth-shape cycle disturbance

The saw-tooth-shape disturbance refers to the disturbance cyclically generated by the processing edge hits whilst cutting the screw in tap processing. The saw-tooth-shape cycle disturbance adopted in the simulation can be expressed as

$$d_x(t) = \begin{cases} 0 & (0 \leq t \leq 0.18, 1.98 < t \leq 2.00) \\ -5\left\{1 + \sin\left(\dfrac{25\pi t}{9}\right)\right\} \\ & (0.36 < t \leq 0.72, 1.08 < t \leq 1.44, 1.80 < t \leq 1.98) \\ -5\left\{1 - \sin\left(\dfrac{25\pi t}{9}\right)\right\} \\ & (0.18 < t \leq 0.36, 0.72 < t \leq 1.08, 1.44 < t \leq 1.80). \end{cases}$$

Its wave is shown in Fig. 7.5.

In order to compare it with the master-slave synchronous positioning control method with the saw-tooth-shape cycle disturbance, the simulation results of the tracking control method between the servo system without position synchronization is shown in Fig. 7.6. The trajectory error $e(t) = p_x(t) - p_y(t)$ of axis y to axis x is only shown, which is different from the simulation results with step disturbance.

Fig. (b) has almost the same results when existing step disturbance. From the Fig. (c) and the results based on Fig. (a), axis y can be synchronized on position with axis x when exhibiting the saw-tooth-shape cycle disturbance. However, from the graph of trajectory error, there are two times of trajectory error 0.3[mm] in Fig. (c) comparing with 0.15[mm] in Fig. (a) when considering the impact of the saw-tooth-shape cycle disturbance input between

7.1 The Master-Slave Synchronous Positioning Control Method 157

0~2[s]. If there are no saw-tooth-shape cycle disturbances between 2~4[s], the results are consistent with the situation of step disturbance. Therefore, the effectiveness of the master-slave synchronous positioning control method was verified.

(2) Experiment of Master-Slave Synchronous Positioning Control

In the former part, a simulation was made with a disturbance generated in the computer and good results were obtained. Next, an experiment will be made with the actual XY table. The experiment is carried out with two input methods of disturbance $D_x(s)$. One is with a disturbance generated in the computer, i.e., disturbance is supposed to exist in the controller of the XY table. The disturbance is put into the computer and it is generated considering the various input possibilities of the actual equipment. Another one is that the disturbance is put physically into the actual experiment equipment and it is generated according to the actual situation of operation.

(i) In the case of putting the disturbance into the computer

With the same input command as the former part, an experiment is carried out with the same conditions. Fig. 7.7 illustrates the experimental results under the step disturbance with the master-slave synchronous positioning control method and simulation results of the tracking control method between two servo system without position synchronization. Fig. 7.8 illustrates the

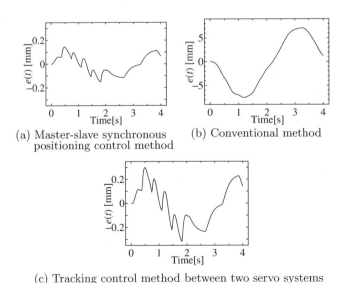

Fig. 7.6. Simulation results with saw tooth state cycle disturbance

158   7 Master-Slave Synchronous Positioning Control

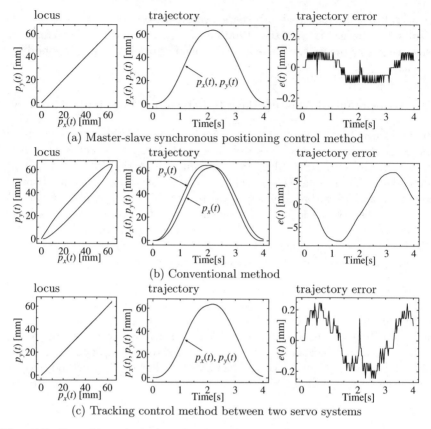

Fig. 7.7. Experimental results of master-slave synchronous positioning control method for the step disturbance based on XY table

experimental results with the saw-tooth-shape cycle disturbance under the same conditions.

The whole simulation results and experimental results are consistent. However, from the trajectory error with the saw-tooth-shape cycle disturbance, the saw-tooth-shape cycle disturbance can be shown in the simulation. But it cannot be shown in the experiment. The reason is that the impact of quantization error is almost unchanged in order to connect the A/D, D/A converter into the controller between the XY table and the personal computer.

(ii) Disturbance input in the actual equipment

In order to approach the motion conditions adopted in the actual equipment, an experiment, in which disturbance was directly added to the experiment equipment, was carried out. In the XY table, the y axis is completely moved along the direction of the x axis in order to make the y axis moving based on the x axis. Therefore, it is only possible to put disturbance into the x

7.1 The Master-Slave Synchronous Positioning Control Method   159

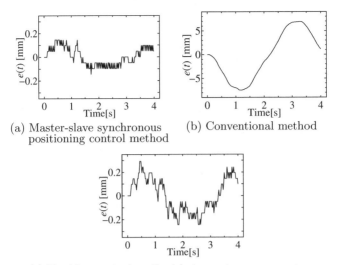

Fig. 7.8. Simulation results for the saw tooth state cycle disturbance wave-like based on XY table

axis because the force opposite to the movement of the x axis can be added in the y axis. When the XY table is moving, a step disturbance is generated due to the external force added on the y axis of 700~800[N] in the XY axis. In the XY table, an experiment was carried out based on the tracking control method between two servo systems and the master-slave synchronous positioning control method. The input command is

$$u_x(t) = 28.2 \qquad (0 \le t \le 5).$$

When adding disturbance to the actual equipment, the experimental results based on the master-slave synchronous positioning control method and tracking control method between two servo systems are illustrated in the Fig. 7.9, respectively. The left-hands side illustrates the results of the XY table locus, trajectory error $e(t) = p_x(t) - p_y(t)$ of the y axis corresponding to the x axis. From the locus of the XY table, position synchronization of the y axis output in the x axis output can be realized based on both methods. However, from the trajectory error graph, there appeared large errors about 0.6[mm] after the beginning of the experiment the both methods. The reason is that the feedback signal is not input during the initial step in order to discretely approximate the differential of feedback signal of the x axis position output for both methods. In Fig. 7.9(c), the error reduction is very slow after the beginning of the experiment and the maximal error amplitude is 0.15[mm] after dropping of the constant. But the amplitude of the constant error in Fig. (a) is very small at 0.07[mm]. Therefore, the effectiveness of the master-slave syn-

160   7 Master-Slave Synchronous Positioning Control

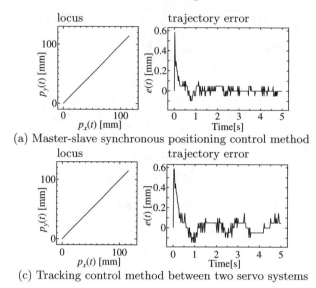

Fig. 7.9. Experimental results with actual disturbance based on XY table

chronous positioning control method when adding disturbance to the actual equipment was verified.

7.2 Contour Control with Master-Slave Synchronous Positioning

In the proposed master-slave position synchronization, there exists a large deviation in the following locus of objective locus at the corner of the trajectory when using contour control for any objective trajectory.

In master-slave position synchronization, the response locus of the objective trajectory has a deviation because of the response delay to the objective trajectory at the corner of the trajectory. Therefore, the following two control methods added to the master-slave synchronous positioning control method are proposed, namely, the method with command of extending the linear interval before following a locus approximating the objective trajectory after putting into objective trajectory of the position, and the method of high-precision contour control with a little time synchronization at the corner.

The servo controllers of industrial mechatronic systems almost all control each axis independently. In the proposed method, it is not necessary to change the existing hardware and software, which provide commands, it is only necessary to revise with simple definition of the master axis and the slave axis.

7.2.1 Derivation of the Contour Control Method with Master-Slave Synchronous Positioning

(1) Definition of the Problem

The control objective is a kind of mechatronic systems which has a structure with two axes like an XY table. The control purpose is to realize high-precision contour control of a mechatronic servo system tracing an objective trajectory even without strict property (K_p). Moreover, for the mechatronic servo system with a defined master axis and slave axis, if the problem for the maximal two axes could be solved, it can be expanded to the multiple axes if the mechatronic servo system with multiple axes also contains the same relationship between the master axis and the slave axis. In the typical processing as tap, since the impact of disturbance mixed into the control system on the slave axis can be neglected comparing with that on the master axis in the direction of rotation, therefore, a disturbance can be only mixed into the master axis.

In the contour control of the mechatronic servo system, the objective locus is approximated by different lines (refer to 1.1.2 item 8). In the nth linear interval, the velocity v is put into move from the objective point (x_n, y_n) as original point to the $n+1$th objective point (x_{n+1}, y_{n+1}). If the control objective reaches the objective point (x_{n+1}, y_{n+1}), the same movement will begin from this new original point. Such a kind of movement will stop until when reaching the final objective point. However, there exists a delay in the servo system and the final part of the line trajectory is lost because the position output cannot reach the objective point even when the position input of the objective locus reaching the objective point. Therefore, contour control for the trajectory composed of the line trajectories will be separately considered into a linear interval and a corner part.

(2) Control Method with a Linear Interval

In order to realize the correct contour control, it is necessary to make the proportional relationship between the master-axis position output $P_x(s)$ and the slave-axis position output $P_y(s)$. The control system with this relation adapts the master-slave synchronous positioning control method introduced in 7.1. In the master axis, the velocity input $U_x(s)$ as standard is the input and in the slave axis, the master-axis position output is regarded as the inverse dynamics modification element $F_s(s)$ of the slave axis. The line between the nth objective point (x_n, y_n) and the $n+1$ objective point (x_{n+1}, y_{n+1}) is given after multiplying the coefficient k_c (region A in Fig. 7.10). This master-axis position $P_x(s)$ and slave-axis position $P_y(s)$ is expressed according to the 1st order model of the servo system as

$$P_x(s) = \frac{K_{px}}{s + K_{px}} \frac{1}{s} U_x(s) + \frac{1}{s + K_{px}} D_x(s) \qquad (7.9a)$$

$$P_y(s) = \frac{K_{py}}{s + K_{py}} k_c F_s(s) P_x(s) \qquad (7.9b)$$

where $D_x(s)$ denotes velocity disturbance. K_{px} and K_{py} have the meanings of K_{p1} in equation (2.20) in the low speed 1st order model of 2.2.3 about the master axis and the slave axis, respectively. Moreover, $F_s(s)$ is as the following equation when using inverse dynamics based on the slave-axis feature.

$$F_s(s) = \frac{s + K_{py}}{K_{py}}. \tag{7.10}$$

Therefore, in the contour control in this region, the master-axis position input command is $u_x(t) = \mathcal{L}^{-1}\{U_x(s)/s\}$ and the slave-axis revised input command is $u_y(t) = \mathcal{L}^{-1}\{k_c F_s(s) P_x(s)\}$. They are given to the servo system as the command of each sampling time interval Δt_p.

(3) Control Method with the Corner Part

Since there is a response delay corresponding to the objective trajectory, the response locus will be missed to the objective locus in the corner part. In order to prevent locus deterioration due to such a miss, after the position input reaching the objective point and at the moment of the following locus reaching into the distance within the time of $v\Delta t_p$ from the objective point (region B in Fig. 7.10), the control method (Fig. 7.1) on the linear interval will be continued without a change of k_c and the command time will be also lasted. The input scale of extended command is within the radius of $v\Delta t_p$ from objective point and until realizing position output. The reason is that the advancing distance within one sampling time interval Δt_p with the given objective tangent velocity v is $v\Delta t_p$. When the following locus reaches within $v\Delta t_p$, in order to change the position input command into the next linear interval position input, v is as the velocity introduced in 7.2.1(1) generally, even generating locus deviation deterioration as Fig. 7.10. Since Δt_p is also very small, the error is actually very slight. Moreover, proper v and Δt_p can

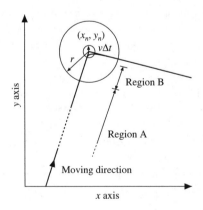

Fig. 7.10. Contour control at the corner part

be previously worked out according to the allowable error. After the following locus reaches the radius $v\Delta t_p$ of the objective point and based on Fig. 7.1, the next control of the linear interval will be carried out with a command for synchronization. The above proposal is the contour control method of master-slave synchronous positioning.

7.2.2 Property Analysis and Evaluation of the Contour Control Method with Master-Slave Synchronous Positioning

(1) Property Analysis of the Contour Control Method with Master-Slave Synchronous Positioning

The effectiveness of the contour control method of master-slave synchronous positioning is evaluated according to the analytical solution using a mathematical model of equation (7.9a)~(7.10). In the equation (7.9a), (7.9b), a inverse Laplace transform (refer to appendix A.1) is conducted with $U_x(s) = v_x/s$, $D_x(s) = d_x/s$ as

$$p_x(t) = x(0)e^{-K_{px}t} + \left(\frac{e^{-K_{px}t} - 1}{K_{px}} + t\right)v_x + \frac{1 - e^{-K_{px}t}}{K_{px}} d_x \quad (7.11a)$$

$$p_y(t) = y(0)e^{-K_{py}t} + k_c x(0)e^{-K_{px}t} + k_c \left(\frac{e^{-K_{px}t} - 1}{K_{px}} + t\right)v_x$$

$$+ k_c \frac{1 - e^{-K_{px}t}}{K_{px}} d_x \quad (7.11b)$$

where v_x denotes the velocity of the x direction if the objective tangent velocity is v. d_x denotes disturbance.

Using equation (7.11a), (7.11b), locus error in the vertical direction of the following locus corresponding to the objective locus in the contour control method of the master-slave synchronous positioning can be calculated. Locus error can be calculated according to $|p_x(t)\sin\varphi - p_y(t)\cos\varphi|$ if angel between the x axis and the objective locus is calculated with φ. Then it is put into equation (7.11a), (7.11b). And here, proportional constant k_c is changed as $\tan\varphi$ by using φ, and based on the handling in the corner part with the contour control method of master-slave synchronous positioning, the small values of $x(0)$, $y(0)$ as initial values for the next interval can be approximated to $x(0) = y(0) = 0$ because the following locus is made to approximate the objective point. According to above procedure,

$$e(t) = |\alpha(\sin\varphi - \tan\varphi\cos\varphi)| = 0 \quad (7.12)$$

in theory, the locus error will be 0 when mixed with any kinds of disturbances. But there exist as

$$\alpha = \left(\frac{e^{-K_{px}t} - 1}{K_{px}} + t\right)v_x + \frac{1 - e^{-K_{px}t}}{K_{px}} d_x. \quad (7.13)$$

Namely, if there are no modeling errors in the contour control method of master-slave synchronous positioning and the slave axis is tracing correctly the master axis at any time, it shows that the locus error of contour control is 0.

(2) Property Analysis of the Modeling Error

In the contour control method of master-slave synchronous positioning, the servo system is expressed by the 1st order model. The position synchronization is carried out using its inverse dynamics equation (7.10). In fact, it is very difficult to make the property values K_{px}, K_{py} of each axis consistence completely because of the variation of moment of inertial according to the mechanical movement states and the variation of the spring constant. For example, K_{px}, K_{py} are not consistent even thought that disturbance is not mixed. Therefore, the deterioration occurred in the contour control performance of the contour control method of the master-slave synchronous positioning because there exist modeling errors in the K_{py} of the mathematical model. The deterioration degree, i.e., robustness of this contour control method corresponding to the modeling error K_{py}, is discussed. When K_{py} of the modification element $F_s(s)$ is different from ΔK_{py} of the actual control objective, the relationship between the proposed method and modeling error can be distinguished by investigating the control performance of the master-slave synchronous positioning control method. And here, the property is investigated when K_{py} and ΔK_{py} are different from the previous assumption of gain of the modification element $F_s(s)$ and disturbance and the initial value is 0. the modification element $\hat{F}_s(s)$ is expressed if existing modeling error K_{py} as

$$\hat{F}_s(s) = \frac{s + K_{py} + \Delta K_{py}}{K_{py} + \Delta K_{py}}. \tag{7.14}$$

The stationary term of locus error $e(t)$ using equation (7.14) is expressed as below when $F_s(s)$ in equation (7.11a), (7.9b) are changed into $\hat{F}_s(s)$ and deviated analytical solution $p_y(t)$ is used.

$$e = \left| \frac{\Delta K_{py}}{K_{py}^2 + K_{py}\Delta K_{py}} v_x \sin \varphi \right| \tag{7.15}$$

where locus error $e(t)$ is as e which is not changed depends on time t. This equation (7.15) expresses the locus error of contour control when the adopted contour control method of the master-slave synchronous positioning with modeling error ΔK_{py}.

If existing modeling error, the significance of using the contour control method of the master-slave synchronous positioning is evaluated according to the comparison with contour control performance without complete position synchronization. In the conventional method without position synchronization, the locus error is expressed as below if $F_s(s)P_x(s)$ in equation (7.11a)

7.2 Contour Control with Master-Slave Synchronous Positioning

and (7.9b) is changed into $U_x(s)/s$ and the stationary item can be calculated if put into the analytical solution $p_y(t)$

$$e = \left| \frac{K_{px} - K_{py}}{K_{px}K_{py}} v_x \sin\varphi \right|. \tag{7.16}$$

This equation (7.16) is also about e which is not dependent on time t. If ΔK_{py} in equation (7.15) is changed, the locus error will be adjusted in the small scale comparing with the locus error in equation (7.16) of conventional method. And here, $K_{py} = n_{xy}K_{px}$ when carrying out analysis. The solution of inequality of the conventional method and the contour control method of master-slave synchronous positioning are as

$$(1 - n_{xy})K_{px} \leq \Delta K_{py} < \infty, \quad (2 \leq n_{xy}) \tag{7.17a}$$

$$(1 - n_{xy})K_{px} \leq \Delta K_{py} < \frac{n_{xy}(1 - n_{xy})}{n_{xy} - 2} K_{px}, \quad (1 < n_{xy} < 2) \tag{7.17b}$$

$$\frac{n_{xy}(1 - n_{xy})}{n_{xy} - 2} K_{px} < \Delta K_{py} < (1 - n_{xy})K_{px}, \quad (0 < n_{xy} < 1). \tag{7.17c}$$

When the x axis is the master axis and the y axis is the slave axis, generally, in order to define the response property of the slave axis faster than that of the master axis and $K_{px} \leq K_{py}$, the results in the scale of $n_{xy} \geq 1$ in equation (7.17a), (7.17b) is very important.

In order to evaluate the appropriation, when these condition equations are regarded as evaluation criteria of the contour control method of master-slave synchronous positioning, a simulation of contour control is conducted. With conditions of $\Delta t_p = 10[\text{ms}]$, $v = 10[\text{mm/s}]$, $K_{px} = 5[1/\text{s}]$, the results with five types of K_{py}, 7, 10, 20, 30, 50[1/s], are illustrated in Fig. 7.11. The objective locus is performed $(0,0) \to (20,20)$ as the objective point. The horizontal axis is the ratio $R_y = (K_{py} + \Delta K_{py})/K_{py}$ of modeling error corresponding

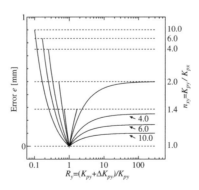

Fig. 7.11. The relationship between the locus error e and the modeling error rate R_y

to the actual value K_{py}. The left vertical axis is the locus error e. The right vertical axis is the ratio of $n_{xy} = K_{py}/K_{px}$. In these figures, the dotted line in the graph is the locus error of the contour control if there is no position synchronization of the conventional method. If the locus error of the contour control method of the master-slave synchronous positioning is under the dotted line, the significant of position synchronization can be judged even there have been modeling errors. Furthermore, the effective scale of the proposed method can be shown in the scale of the horizontal axis R_y described by the unbroken line in the graph.

From these results, the scale, when the locus error of contour control method of master-slave synchronous positioning with $R_y < 1$ is bigger than that of the conventional method, exists. Therefore, the proposed method is effective within the limit scale in which $R_y > 1$, $n_{xy} \geq 2$, and the locus error is smaller than the conventional method with any ΔK_{py}, and $1 < n_{xy} < 2$. However, in fact, if the property of the servo system of the equipment is known clearly, the modeling error of K_{py}, in general, will be several percents to times of percents. From the results of Fig. 7.11, the locus error is very slight when $R_y = 1$ is equivalent to the actual modeling error. Based on the above analytical results of the modeling error and considering that the current state of the actual modeling error, in fact, it is possible to adapt effectively the contour control with position synchronization of the contour control method of the master-slave synchronous positioning.

7.2.3 Experimental Test of the Contour Control Method of Master-Slave Synchronous Positioning

The experiment with the contour control method of master-slave synchronous positioning was carried out using an actual XY table (refer to E.4 about experiment equipment). The experimental conditions are, Δt_p=10[ms], v=3(about 1/23 rated speed)[mm/s], the position loop gains of the 1st order model are K_{px}=5[1/s] of master axis and K_{py}=10[1/s] of the slave axis. The objective trajectory, illustrated in Fig. 7.12, is moved from 1 at $(0,0)$ to 2 at $\rightarrow$ to 3 at $\rightarrow$ to 4 at $\rightarrow$ to 1. The disturbances are put into as -5[mm/s] step between 3~7 second at the region of 2→3 and as -1[mm/s] step between 3~7 second at the region of 3→4. Fig. 7.12(a) is about the conventional method by which each axis is controlled independently. Fig. (b) is about the contour control method of the master-slave synchronous positioning. Fig. (c) illustrates the results of the contour control method of master-slave synchronous positioning when adding the modeling error $\Delta K_{py} = 1$[1/s]. The left graph shows the locus if the horizontal axis is the position output of the x axis and the vertical axis is the position output of the y axis. The right graph shows the time change of the locus error. In Fig. (a), the error occurred with an average 0.1[mm] according to the difference of servo features among axis (K_{px}=5, K_{py}=10) and also occurred due to disturbance influence. Thus, the following locus cannot reach the final objective point. In Fig. (b), in all regions, the error is about 0.02[mm] if contour control is correctly performed because of good position synchronization even in the part added with disturbance. In addition, in the Fig. (c), the locus error is 0.03[mm] when the precision of contour control is very high.

In order to use the differential of the modification element of the contour control method of master-slave synchronous positioning, the noise mixed into the input signal of the slave axis should be considered. If making differential (discrete) on a signal with much noise, the problem of amplitude increment will be generated. If using pulse output of the encoder which is often applied in the mechatronic servo system of the industrial field, there will be no problem on the discrete of master-axis output values. Even using the tachogenerator output as position output and these values including the integral value of the tachogenerator output, there are also no problems on their discrete. In the current experiment, good results were obtained when using the tachogenerator output.

Based on the proposed position synchronization contour control method, good control performance can be verified not only by theoretical analysis but also with experiment results.

7 Master-Slave Synchronous Positioning Control

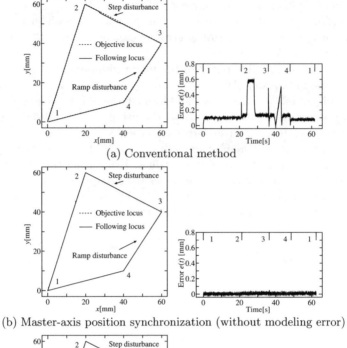

(a) Conventional method

(b) Master-axis position synchronization (without modeling error)

(c) Master-axis position synchronization (with modeling error)

Fig. 7.12. Experimental results of contour control based on XY table

Glossary

A. Definition of device

axis: minimal unit of mechanism corresponding to one motor movement, such as rotational axis, proceeding axis, etc., including control part of each motor in each articulated part of robot arm.

current control part: control current flowing in motor by current command of motor and current sensor.

command part: generation part of position and velocity (trajectory) for the mechatronic servo system movement driven by numerical input or taught data. (management part) reference input generation part.

reference input time interval: time interval of data alternation of objective trajectory given from reference input generation part to position control part, also called as system clock.

encoder resolution: position detector installed in motor axis, (using PPR: pulse per resolution).

flexible arm: spring factor included in deceleration part of robot arm is with low rigid, robot arm is flexibly structured, and it is needed in the flexible body of arm but not rigid. It is the name of arm with flexible body.

management part: management of movement of robot arm, movement related with surrounding devices (command of movement, emergence stop, etc. according to operation procedure).

mechanism: construct the mechanic movement part of mechatronic servo system.

mechatronic servo system, mechatronic system: The overall title of servo part of mechatronic machine composed by (management part) reference input generator, position control part, velocity control part, current control part, power amplifier, motor and mechanism.

motor: drive mechanism transforming power into rotational movement and change torque of rotational movement according to input current.

objective point: angle part when approximating linearly the given objective locus in contour control.

position control system, servo, servo system: from position command of one axis of mechatronic servo system, to position output of motor, including position control part, velocity control part, current control part, power amplifier part, motor part and mechanism part.

power amplifier: amplifier of providing power to motor according to current command, generally, using amplifier with PWM (pulse width modulation) pattern.

position control part: control motor position by given position information of motor and observer value, generally, used for ratio control.

reduction ratio: ratio of input rotation times, of decelerator connecting motor and mechanism part, and output rotation times.

reference input generator: compute something in each designated time interval for position control of each link based on objective trajectory and put into position control part.

sampling time: time interval till exporting torque (velocity) command, by observing state based on the (updated) data input when velocity control part (position control part) is structured with digital controller, and necessary computing.

servo amplifier: power amplifier part providing output rated with input signal (current control part).

servo controller: from position input of one axis of mechatronic servo system to current input of motor, including (position control part,) velocity control part, current control part and power amplifier part.

servo parameters: parameters including in servo controller, such as K_p, K_v, etc.

torque resolution: determined from the motor current whose analogue current command calculated in current control part is transformed by D/A transformation, and feedback into current control part is determined by A/D transformation resolution to be driving force of motor.

velocity control part: control velocity of motor by velocity command of motor and observer value, generally used for ratio integral control.

velocity control part: from velocity input of one axis of mechatronic servo system, to velocity output of motor, including velocity control part, current control part, power amplifier and motor part.

B. Definition of control method

contouring control: control to realize trajectory of position output of mechatronic servo system. In industrial application, the curve of objective locus is approximated by line and the tangential velocity is always constant.

full-closed loop: it is the control system structured by feedback system with the information of each moveable tip or motion tip.

modified taught data method: method for improving the contour control performance through the delay dynamics compensation of mechatronic servo system.

position control: for realizing position of mechatronic servo system without problems in the part of locus. It is called PTP(point to point).

semi-closed loop: it is not constructed by the feedback system with the information of each moveable tip or motion tip, but by the information of servo actuator part.

synchronous position control: master and slave: realize synchronous position control through the compensation of delay dynamics.

C. Definition of model

1st order model: model expressed by 1st order model of the overall mechatronic servo system. In general, it can be adopted with the velocity below 1/20 motor rated velocity.

2nd order model: model expressed by 2nd order system of the overall mechatronic servo system. In general, it can be adopted with the velocity of $1/20 \sim 1/5$ motor rated velocity.

4th order model: model with the combination that motor rotation part and mechanism part of mechatronic servo system are expressed by two mass model, and electric part of servo controller is expressed by 2nd order model.

centrifugal force: inertial force of object in rotational coordinate system. Namely, assumed force of movement with acceleration faced to rotation center by taking into account the halt balance.

coriolis force: act on the object moving in the rotational coordinate system and the force being proportional with velocity.

linear dynamic model in joint coordinates: model constructed by linear approximation of dynamics of articulated robot arm independently in each coordinate axis in joint coordinates.

linear dynamic model in working (cartesian) coordinates: model constructed by linear approximation of dynamics of articulated robot arm independently in each coordinate axis in cartesian (orthogonal) coordinate.

modeling error: error of modeling control object with the actual control object.

reduced order model: General title of 1st and 2nd order model of mechatronic servo system.

two mass model: model constructed by connecting the system model, structured by the mechanism driven using motor, with motor inertia moment and load inertia moment with spring.

D. Definition of performance

locus: final results of position of mechatronic servo system without concerning time shift , and not time function.

locus error: error between objective locus and tracing locus, not time function.

locus error: phenomenon of error occurred in locus because of periodic ripple of motor position.

trajectory: time shift of position of mechatronic servo system, and time function.

trajectory error: error between objective trajectory and tracing trajectory, time function.

torque saturation: phenomenon can not be permitted in command that motor torque output reaching the limitation of output about performance of current control part or power amplifier part.

velocity ripple: phenomenon of periodic ripple of motor velocity.

Nomenclature

The symbols of each chapter in this book are summarized.

2. Mathematical model of a mechatronic servo system
2.1 4th order model of one axis in a mechatronic servo system

Symbols	Units	Meanings
$u(t)$	rad	angular input of motor
$y(t)$	rad	angular output of motor
J_M	kgms^2	inertial moment of motor
J_L	kgms^2	inertial moment of mechanism
D_L	Nms	viscous coefficient of mechanism part
K_L	Nm	spring constant of mechanism part
ω_L	rad/s	natural angular frequency of mechanism part
ζ_L		damping factor in mechanism part
T_M	Nm	torque occurred in motor
T_L	Nm	reaction force from mechanism part
N_G		gear ratio
N_L		moment ratio ($J_L/N_G^2 J_M$)
θ_M	rad	angular output of motor
θ_L	rad	angular output of mechanism part
K_p	1/s	position loop gain
K_v	1/s	velocity loop gain
J_T	kgms^2	total inertial moment of motor and mechanism part
K_v^g		velocity amplifier gain
c_p		normalized position loop gain
c_v		normalized velocity loop gain

174   Nomenclature

2.2 Reduced order model of one axis in a mechatronic servo system

Symbols	Units	Meanings
$G_{c1}(s)$		normalized low speed 1st order model
$G_{c2}(s)$		normalized middle speed 2nd order model
K_{p1}	1/s	position loop gain of low speed 1st order model
K_{p2}	1/s	position loop gain of middle speed 2nd order model
K_{v2}	1/s	velocity loop gain of middle speed 2nd order model
c_{p1}	1/s	position loop gain of normalized low speed 1st order model
c_{p2}	1/s	position loop gain of normalized middle speed 2nd order model
c_{v2}	1/s	velocity loop gain of normalized 2nd order model

2.3 Linear model of the working coordinates of an articulated robot arm

Symbols	Units	Meanings
θ_1	rad	Angle of 1st axis
θ_2	rad	angle of 2nd axis
p_x	m	position of X axis
p_y	m	position of Y axis
l_1	m	length of 1st axis
l_2	m	length of 2nd axis
ΔT	s	reference input time interval
$\lambda(t)$	s	$\lambda(t) = t + (e^{K_p t} - 1)/K_p$
$\hat{p}_x$	m	position of X axis in working coordinate model
$\hat{p}_y$	m	position of Y axis in working linear model
v_1	rad/s	velocity of 1st axis
v_2	rad/s	velocity of 2nd axis
v_x	m/s	velocity of X axis
v_y	m/s	velocity of Y axis
ϵ_x	m/s	error at the X direction of working linear model
ϵ_y	m/s	error at the Y direction of working linear model
v	m/s	objective velocity

3. Discrete time interval of a mechatronic servo system
3.1 Sampling time interval

Symbols	Units	Meanings
$G_1(s)$		transfer function of 1st order system
$G_{L1}(s)$		transfer function of 1st order system with time delay
$G_{P1}(s)$		transfer function of 1st order system with time delay Pade approximation
f_c^0	Hz	Sampling frequency
L_1	s	sum of required time from state sample of position loop to control input calculation and delay time in 0th order hold of control input
f_{cP}	Hz	cut-off frequency of transfer function of 1st order system with time delay Pade approximation
f_{c1}	Hz	cut-off frequency of transfer function of 1st order system
Δt_p	s	sampling time interval
q_1		$q_1 = qL_1/\Delta t_p$
f_s	Hz	$f_s = 1/\Delta t_p$

3.2 Relation between reference input time interval and velocity fluctuation

Symbols	Units	Meanings
p_1^s	1/s	pole of 2nd order model
p_2^s	1/s	pole of 2nd order model
e_v^s		maximal constant velocity fluctuation
e_v^t		maximal transient velocity fluctuation
h_r		0th order hold in reference input generator
u_p		position command
h_p		0th order hold in position command part
u_v		velocity command
r		objective trajectory
v_{ref}		objective velocity
n_{pv}		$n_{pv} = K_v/K_p$ gain ratio

3.3 Relation between reference input time interval and locus irregularity

Symbols	Units	Meanings
K_{px}	1/s	position loop gain of x axis of 1st order model
K_{py}	1/s	position loop gain of y axis of 1st order model
J		Jacobian matrix
p_z	m	position of z axis
K_{pz}	1/s	position loop gain of z axis of 1st order model

4. Quantization error of a mechatronic servo system

4.1 Encoder resolution

Symbols	Units	Meanings
ΔN	rev/min	amplitude of velocity fluctuation
f_r	s	frequency of velocity fluctuation
R_E	pulse/rev	encoder resolution
$N_{\max}$	pulse/s	maximal velocity of servo motor
R_N		$R_N = \Delta N/N_{\max}$ velocity fluctuation rate
V_{ref}	pulse/s	command velocity

4.2 Torque resolution

Symbols	Units	Meanings
P_{ref}	pulse	objective position
E_p^s	pulse	position decision error
Δt_v	s	sampling time interval of velocity loop
T_d	s	time of angular velocity output below objective velocity
T_u	s	time of angular velocity output over objective velocity
V_d	pulse/s	velocity of angular velocity output below objective velocity
V_u	pulse/s	velocity of angular velocity output over objective velocity
E_d	pulse	maximal position deviation of angular velocity output below objective velocity
E_u	pulse	maximal position deviation of angular velocity output over objective velocity
T_f	s	period of fluctuation
E_p^r	pulse	amplitude of position fluctuation
E_v^r	pulse/s	amplitude of velocity fluctuation
R_p	pulse/s^2	angular acceleration resolution upper boundary satisfying amplitude condition of position output error
R_v	pulse/s^2	angular acceleration resolution upper boundary satisfying amplitude condition of angular velocity output error
R_A	pulse/s^2	angular acceleration resolution
R_T	Nm	torque resolution
$T_{\max}$	Nm	maximal torque
B	bit	bit number corresponding to torque resolution

5. Torque saturation of a mechatronic servo system

5.1 Measurement method for the torque saturation property

Symbols	Units	Meanings
a	m/s^2	input acceleration
t_M	s	moment of torque taking maximal output
$sat(x)$		saturation curve

Nomenclature

5.2 Contour control method with avoidance of torque saturation

Symbols	Units	Meanings
A_{max}	m/s^2	maximal acceleration
ϵ	m	working precision
V	m/s	command tangential velocity
$r_x(t)$	m	objective trajectory at the direction of x axis
$r_y(t)$	m	objective trajectory at the direction of y axis
$w_x(t)$	m	input considering working precision at the direction of x axis
$w_y(t)$	m	input considering working precision at the direction of y axis
$u_x(t)$	m	revised input at the direction of x axis
$u_y(t)$	m	revised input at the direction of y axis
r_{min}	m	$r_{min} = V^2/A_{max}$ possible minimal radius of circular trajectory of movement for maximal acceleration
r	m	circular radius satisfying working precision ϵ
V_m	m	$V_m = \sqrt{A_{max} r}$ velocity satisfying maximal acceleration A_{max} when drawing radius r
a_{max}	m/s^2	maximal angular acceleration
θ_{c1}	rad	angle with x axis of objective locus 1
θ_{c2}	rad	angle with x axis of objective locus
$F(s)$		modification term

6. The modified taught data method

6.1 Modified taught data method using a mathematical model

Symbols	Units	Meanings
$r(t)$	m	objective trajectory
$G_1(s)$		transfer function of 1st order model
$F_1(s)$		modification term based on 1st order model
$G_2(s)$		transfer function of 2nd order model
$F_2(s)$		modification term based on 2nd order model
ω_c	rad/s	cut-off frequency
γ	1/s	pole of pole assignment regulator by 1st order model
K_s		feedback gain of pole assignment regulator
ϕ_m	rad	maximal phase-lead value of modification term
ω_m	rad	frequency of maximal phase-lead value of modification term
γ_i	1/s	pole of pole assignment regulator by 2nd order model
μ	1/s	pole of observer by 2nd order model

6.2 Modified taught data method by using a Gaussian network

Symbols	Units	Meanings
$\phi(\boldsymbol{x})$		output of Gaussian network
w_i		weight of Gaussian network
$\psi_i(x_i)$		Gaussian unit
M		number of Gaussian units
x_i		input to Gaussian network
m_i		mean of Gaussian unit
σ_i		Variance of Gaussian unit
$x_{\max}$		linear approximation region of Gaussian network
$x_{\max}^p$	m	constant determining linear approximation region of position
$x_{\max}^v$	m/s	constant determining linear approximation region of velocity
$x_{\max}^a$	m/s^2	constant determining linear approximation region of acceleration
E_{rms}		lose function of learning of Gaussian network
E^l		factors of lose function of learning of Gaussian network
$(u^k, \boldsymbol{x}^k)$		taught data for learning of Gaussian network
K		number of taught data
$\boldsymbol{p}$		parameter of Gaussian network
$\boldsymbol{p}_i^{\mathrm{new}}$		modified parameter of Gaussian network
$\boldsymbol{p}_i^{\mathrm{old}}$		parameter of Gaussian network before modification
η		learning rate of Gaussian network

6.3 Modified taught data method for a flexible mechanism

Symbols	Units	Meanings
$R(s)$		objective trajectory
$U(s)$		taught data
$Z(s)$		position of fulcrum of arm
$Y(s)$		output
$G_3(s)$		overall transfer function of control system
$F_3(s)$		modification term

7. Master-slave synchronous positioning control

7.1 The master-slave synchronous positioning control method

Symbols	Units	Meanings
k_c		sloping ratio between two axes of objective trajectory

7.2 Contour control with master-slave synchronous positioning

Symbols	Units	Meanings
v_x	s	velocity input to master axis(x axis)
e	mm	locus error
$\hat{F}(s)$		modification term if existing modeling error
R_y		$R_y = (K_{py} + \Delta K_{py})/K_{py}$ coefficient for modeling error evaluation

Experimental Equipments

The main experimental device using in the experiment of this book are illustrated.

E.1 DEC-1

DEC-1 (made by Yahata Electric Machinery Inc.) using in section 2.1, 2.2, 3.2, 3.3, 4.1, 5.1, 5.2 is shown in Fig.E.1. Its specifications are given in table E.1. DEC-1 is composed of servo controller, servo motor, coupling as mechanism part as well as load generator. This experimental device is made from the DC servo motor and servo controller used actually in industry. It is equivalent to the driving part or mechanism part adopted in each axis of

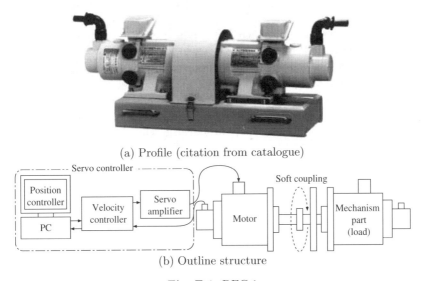

(a) Profile (citation from catalogue)

(b) Outline structure

Fig. E.1. DEC-1

Table E.1. Specification of DEC-1

rated output of motor	kW	0.2
rated torque of motor	kgm	0.195
rated velocity of motor	rev/min	1000
inertia moment of motor axis J_M	kgm^2	0.00224
inertia moment of mechanism part J_L	kgm^2	0.00653
natural angle frequency of mechanism part ω_L	rad/s	94.2
damping rate of mechanism part ζ_L		0.002
encoder resolution	pulse/rev	2000
gear deceleration ratio N_G		1

mechatronic servo system, such as industrial robot, working machine, etc.. If the analysis or control problems of mechatronic servo system using this device can be solved, it is possible to analyze the improvement of control performance of the general industrial mechatronic servo system regulated for having similar properties in each axis, and concrete its improvement strategy. Motor of DEC-1 and load generator are connected by soft-coupling. In this experimental device, velocity control part, current control part, power amplifier part in servo controller are structured by hardware analogue circuit. Position control part is structured by software in computer. Therefore, velocity loop gain K_v is needed to be changed with the regulation of changeable resistance. Position loop gain K_p can be easily changed in software of computer.

E.2 Motoman

The profile of Motoman (made by Yaskawa Inc.) used in section 2.3, 6.1 is shown in Fig.E.2 and its specifications are given in table E.2, respectively. Motoman is an industrial articulated robot arm. Its transportable weigh is from 3 to 150[kg]. It is classified from K3 to K150.

Most of industrial robot arms including Motoman are moved according to the designated taught position series and their velocity. The robot arm using teaching box is moved by taught position and hence its position must be memorized. The taught velocity is given by key input in operation panel. After given all position and velocity, robot arm will move when pushing play key of operation panel.

Fig. E.2. Profile of Motoman (citation from catalogue)

Table E.2. Specification of Motoman K10

(a) Overall specification

degree of freedom		6
precision of repeated PTP control	mm	±0.1
power capacity	kVA	8
transportable weigh	kg	10
body mass	kg	300

(b) Specification of each axis

		1 axis	2 axis	3 axis	4 axis	5 axis	6 axis
length of arm	mm	200	600	770	-	100	-
maximal velocity	rad/s	2.09	2.09	2.09	4.59	4.59	6.98

E.3 Performer MK3S

The profile of Performer MK3s (made by Yahata Electric Machinery Inc.) used in section 5.2 is shown in Fig.E.3 and its specification is given in table E.3. In order to be able to construct controller freely in Performer MK3S, in the authors' laboratory, velocity loop is constructed by hardware in servo controller. Nevertheless, the position loop is rebuilt to be able to construct in computer. Therefore, position loop gain can be set freely in computer.

E.4 XY Table

XY table (made by Yaskawa Inc.) used in section 6.2, 7.1, 7.2 is shown in Fig. E.4 and its specification is given in table E.4. XY table is the device used for transferring knives of working machine because of its independent movement of x axis and y axis according to the ball spring installed in two servo motors, respectively. For making similar of XY table with Performer MK3S, velocity loop is constructed by hardware in servo controller and position loop is constructed in computer. Therefore, position loop gain can be set freely in computer.

Fig. E.3. Profile of Performer MK3S (citation from catalogue)

Table E.3. Specification of Performer MK3S

(a) Overall specification

degree of freedom		5
driving properties	V/pulse	5.0[V]/(2048[pulse/rev]×3000[rpm]/60[s])
detection properties	V/pulse	0.5[V]/(2048[pulse/rev]×1000[rpm]/60[s])
encoder resolution	pulse/rev	8192
transportable mass	kg	2(maximal velocity), 3(low velocity)
body brief mass	kg	32

(b) Specification of each axis

		1 axis	2 axis	3 axis	4 axis	5 axis
length of arm	mm	135	250	215	100	-
output	W	80	80	80	30	30
rated torque	Nm	0.319	0.319	0.159	0.095	0.095
rated rotation number	rpm	2400	2400	3000	3000	3000
rated voltage	V	100	100	100	100	100
rated current	A	2.2	2.2	0.9	0.63	0.63
deceleration ratio		1/120	1/160	1/160	1/120	1/88
inertia moment of motor axis	$\times 10^{-7} \text{Nms}^2$	4.0	4.0	2.7	2.1	2.1

Table E.4. Specification of XY table

rated output of motor	kW	0.2
rated torque of motor	kgm	0.065
rated velocity of motor	rpm	3000
spring pitch	mm	1.4

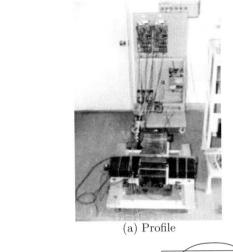

(a) Profile

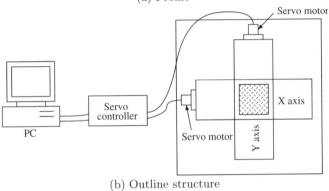

(b) Outline structure

Fig. E.4. XY table

Appendix

A.1 Laplace Transform and Inverse Laplace Transform

If there is a function $f(t)$ on time t

$$\int_0^\infty f(t)e^{-st}dt = \mathcal{L}[f(t)] = F(s) \tag{A.1}$$

it is called as Laplace transform of $f(t)$[36] The inverse transform of equation (A.1)

$$f(t) = \mathcal{L}^{-1}\{F(s)\} \tag{A.2}$$

is called as inverse Laplace transform. In s domain, the inverse Laplace transform of rational function $F(s)$ is transformed by partial fraction factorization as

$$F(s) = \frac{N(s)}{D(s)} = \frac{K_1}{s-s_1} + \frac{K_2}{s-s_2} + \cdots + \frac{K_i}{s-s_i} + \cdots + \frac{K_n}{s-s_n} \tag{A.3}$$

and the determination of its coefficients are calculated by

$$K_i = \left[\frac{N(s)(s-s_i)}{D(s)}\right]_{s=s_i}. \tag{A.4}$$

Therefore, Laplace transform is as illustrated in table A.1.

Table A.1. Laplace transform table

$f(t)$	$F(s)$	$f(t)$	$F(s)$
$\delta(t)$	1	$te^{-\sigma t}$	$1/(s+\sigma)^2$
$u(t)$	$1/s$	$\sin \omega t$	$\omega/(s^2+\omega^2)$
t	$1/s^2$	$\cos \omega t$	$s/(s^2+\omega^2)$
$t^2/2$	$1/s^3$	$e^{-\sigma t}\sin \omega t$	$\omega/((s+\sigma)^2+\omega^2)$
$e^{-\sigma t}$	$1/(s+\sigma)$	$e^{-\sigma t}\cos \omega t$	$s/((s+\sigma)^2+\omega^2)$

Table A.2. Formula of Laplace transform

Linear	$\mathcal{L}[af(t)] = aF(s)$, $\mathcal{L}[f_1(t) + f_2(t)] = F_1(s) + F_2(s)$
Differential	$\mathcal{L}[df(t)/dt] = sF(s) - f(+0)$
Integral	$\mathcal{L}\left[\int f(t)dt\right] = F(s)/s - \int_0^{+0} f(t)dt/s$
Shift in t domain	$\mathcal{L}[f(t-L)] = e^{-sL}F(s)$
Shift in s domain	$\mathcal{L}[f(t)e^{-at}] = F(s+a)$
Final value	$\lim_{t\to\infty} f(t) = \lim_{s\to 0} sF(s)$
Initial value	$\lim_{t\to 0} f(t) = \lim_{s\to\infty} sF(s)$
Convolution	$\mathcal{L}[\int_0^t f_1(\tau)f_2(t-\tau)dt] = F_1(s)F_2(s)$

A.2 Transition Response

If given input $U(s)$ in the factors held in transfer function $G(s)$, the output $Y(s)$ is calculated by

$$Y(s) = G(s)U(s) \tag{A.5}$$

By using inverse Laplace transform, the output $y(t)$ is calculated by

$$y(t) = \mathcal{L}^{-1}[Y(s)] = \mathcal{L}^{-1}[G(s)U(s)] \tag{A.6}$$

This $y(t)$ is called as transition response, namely transitional state before reaching constant state [41] [36] The transition response of basic input, such as impulse response $U(s) = 1$, step response $U(s) = 1/s$, ramp response $U(s) = 1/s^2$, etc., can be calculated by putting these basic inputs into (A.5) and calculating $Y(s)$, and finally can be obtained by inverse Laplace transform.

In this book, 1st order system

$$G(s) = \frac{K_p}{s + K_p} \tag{A.7}$$

is always adopted and its impulse response, step response and ramp response respectively are calculated as

$$g(t) = e^{-K_p t} \tag{A.8}$$

$$f(t) = 1 - e^{-K_p t} \tag{A.9}$$

$$h(t) = t - \frac{1}{K_p}(1 - e^{-K_p t}). \tag{A.10}$$

For 2nd order system with two different real roots

$$G(s) = \frac{K_p K_v}{s^2 + K_v s + K_p K_v} \tag{A.11}$$

the impulse response, step response and ramp response respectively are calculated as

$$g(t) = \frac{s_1 s_2}{s_1 - s_2}(e^{s_1 t} - e^{s_2 t}) \tag{A.12}$$

$$f(t) = 1 + \frac{s_2}{s_1 - s_2}e^{s_1 t} + \frac{s_1}{s_2 - s_1}e^{s_2 t} \tag{A.13}$$

$$h(t) = t - \frac{1}{K_p} + \frac{s_2}{s_1(s_1 - s_2)}e^{s_1 t} + \frac{s_1}{s_2(s_2 - s_1)}e^{s_2 t} \tag{A.14}$$

$$s_1 = \frac{-K_v + \sqrt{K_v^2 - 4K_p K_v}}{2}$$

$$s_2 = \frac{-K_v - \sqrt{K_v^2 - 4K_p K_v}}{2}.$$

A.3 Pole Assignment Regulator

To control object expressed by state space

$$\frac{d\boldsymbol{x}(t)}{dt} = A\boldsymbol{x}(t) + \boldsymbol{b}u(t) \tag{A.15}$$

the control input is selected as

$$u(t) = -\boldsymbol{f}\boldsymbol{x}(t) \tag{A.16}$$

and the control purpose $\boldsymbol{x}(t) \to 0$ $(t \to \infty)$ can be implemented with any initial state $\boldsymbol{x}(0) = \boldsymbol{x}_0$. From equation (A.15) and (A.16), the regulator for setting poles (eigenvalue of $A - \boldsymbol{b}\boldsymbol{f}$) of closed-loop system is called as pole assignment regulator [42].

1. Eigen-equation of A is

$$|sI - A| = s^n + a_{n-1}s^{n-1} + \cdots + a_1 s + a_0 \tag{A.17}$$

and conversion matrix for controllable canonical form is calculated

$$T = (\boldsymbol{b}, A\boldsymbol{b}, \cdots, A^{n-1}\boldsymbol{b}) \begin{pmatrix} a_2 & a_3 & a_4 & \cdots & a_n & 1 \\ a_3 & a_4 & \cdots & a_n & 1 & \vdots \\ a_4 & \cdots & a_n & 1 & & \vdots \\ \vdots & a_n & 1 & & & \vdots \\ a_n & 1 & & & & \vdots \\ 1 & \cdots & \cdots & \cdots & \cdots & 0 \end{pmatrix}. \tag{A.18}$$

2. To pole $\mu_1 \sim \mu_n$

$$(s - \mu_1)(s - \mu_2)\cdots(s - \mu_n) = s^n + d_{n-1}s^{n-1} + \cdots + d_1 s + d_0 \tag{A.19}$$

is calculated.
3. Feedback gain $\boldsymbol{f}$ is calculated by

$$\boldsymbol{f} = (d_0 - a_0, d_1 - a_1, \cdots, d_{n-1} - a_{n-1})T^{-1}. \tag{A.20}$$

A.4 Minimal Order Observer

When calculating state feedback (A.16), the state variable $x(t)$ must be completely observed. However, the state variable actually can not be completely observed. The possible observed variable is

$$y(t) = Cx(t). \tag{A.21}$$

At this time, the structure for estimating the state variable $x(t)$ with observed value $y(t)$ and control input $u(t)$ is called as observer [43]. The observer is implemented in

$$\frac{d\omega(t)}{dt} = \hat{A}\omega + Ky(t) + \hat{B}u(t) \tag{A.22}$$

$$\hat{x}(t) = D\omega(t) + Hy(t) \tag{A.23}$$

by

1. select proper matrix W with $(S \neq 0)$ in

$$S = \begin{bmatrix} C \\ W \end{bmatrix}. \tag{A.24}$$

2. Calculate

$$SAS^{-1} = \begin{bmatrix} A_{11} & A_{12} \\ A_{21} & A_{22} \end{bmatrix}, \quad SB = \begin{bmatrix} B_1 \\ B_2 \end{bmatrix}. \tag{A.25}$$

3. Consider the design parameter

$$\hat{A} = A_{22} - LA_{12} \tag{A.26}$$

the eigenvalues of matrix L are $\gamma_1, \gamma_2, \cdots, \gamma_{n-l}$, and the L can be determined by the pole assignment regulator method.

4. Calculate

$$K = \hat{A} + A_{21} - LA_{11} \tag{A.27}$$

$$\hat{B} = -LB_1 + B_2 \tag{A.28}$$

$$D = S^{-1} \begin{bmatrix} 0 \\ I_{n-l} \end{bmatrix} \tag{A.29}$$

$$H = S^{-1} \begin{bmatrix} I_l \\ L \end{bmatrix}. \tag{A.30}$$

References

The papers included in this book and the literature referred by these papers are summarized in terms of each chapter. The papers being base of each chapter are indicated in parentheses. (in Japanese)

Chapter 2 [1][2][3]
[1] J. Zou, M. Nakamura, S. Goto and N. Kyura: Model Construction and Servo Parameter Determination of Industrial Mechatronic Servo Systems Based on Contour Control Performance, Journal of the Japan Society for Precision Engineering, vol. 64, no. 8, pp. 1158-1164, 1998. (in Japanese)
[2] S. Goto, M. Nakamura and N. Kyura: Reduced Order Model Configuration of Industrial Mechatronic Servo Systems and Its Significance, Proceedings of the 17th SICE Kyushu Branch Annual Conference, 1998. (in Japanese)
[3] S. Goto, M. Nakamura and N. Kyura: Propriety of Linear Model of Servo System for Industrial Articulated Robot Arms and Evaluation of Its Linearization Error, vol. 7, no. 3, pp. 103-112, 1994. (in Japanese)
[4] Y. Fujino and N. Kyura: Motion Control, Sangyo Tosho, pp. 85-117, 1996. (in Japanese)
[5] K. Kuki, T. Murakami and K. Ohnishi: Vibration Control of a 2 Mass Resonant System by the Resonance Ratio Control, Trans. of the Institute of Electrical Engineers of Japan, vol. 113-D, no. 10, pp. 1162-1169, 1993. (in Japanese)
[6] Y. Hori: Introduction to Control of Torsional Vibration System, Proceedings of the IEE of Japan Industrial Applications Society, S. 12-1, 1994. (in Japanese)
[7] S. Arimoto: Dynamics and Control of Robot, Asakura Shoten, 1990. (in Japanese)
[8] Edited by Yaskawa Electric Co.: Basic of Servo Technology for Mechatronics, pp. 12-25, Nikkan Kogyo Shimbun, 1986. (in Japanese)
[9] S. Takagi: Introduction of Analysis Revision 3, Iwanami Shoten, pp. 62-63, 1973. (in Japanese)

Chapter 3 [10][11][12][13][14]
[10] M. Nakamura, H. Koda and N. Kyura: Determination of Required Sampling Rate for Sampling Control of Continuous Contour Control by Servo System, Trans. of SICE, vol. 28, no. 5, pp. 649-651, 1992. (in Japanese)

[11] M. Nakamura, H. Koda and N. Kyura: Determination of Sampling Frequencies for Sampling Control of Servo System with Multi-Samplers, Trans. of SICE, vol. 29, no. 1, pp. 63-70, 1993. (in Japanese)
[12] N. Egashira, M. Nakamura and N. Kyura: Analysis of Locus Ripples at Every Reference Input Time Interval in Mechatronic Servo Systems, Journal of the Robotics Society of Japan, vol. 13, no. 8, pp. 1153-1159, 1995. (in Japanese)
[13] N. Egashira, M. Nakamura and N. Kyura: Analysis of Stational Velocity Ripples at Each Reference Input Time Interval for Mechatronic Servo Systems, Journal of the Robotics Society of Japan, vol. 16, no. 1, pp. 74-79, 1998. (in Japanese)
[14] N. Egashira, M. Nakamura and N. Kyura: Analysis for Transitional Velocity Ripples of Mechatronic Servo Systems at Each Reference Input Time Interval, Trans. of SICE, vol. 34, no. 10, pp. 1504-1506, 1998. (in Japanese)
[15] T. Mita: Design of Digital Control Systems with Operation Time, Journal of SICE, vol. 22, no. 7, pp. 614-619, 1983. (in Japanese)
[16] T. Matsuo: Zeroes and Their Relevance to Control–IV –Relationship between Zeros and Output Responses–, Journal of the SICE, vol. 29, no. 6, pp. 543-550, 1990. (in Japanese)
[17] N. Kyura: Servo Technology –Relationship between Position Loop and Velocity Loop–, Nikkei Mechanical, vol. 226, no. 8, pp.135-140, 1986. (in Japanese)
[18] N. Sasaki: Software of Digital Servo, Kindai Tosho, pp.118-124, 1994. (in Japanese)

Chapter 4 [19][20]
[19] S. Goto, M. Nakamura and N. Kyura: Relationship between Control Performance and Encoder Resolution in Mechatronic Software Servo Systems, Proceedings of the 15th SICE Kyushu Branch Annual Conference, pp. 387-390, 1996. (in Japanese)
[20] S. Goto, M. Nakamura and N. Kyura: Determination Method of Torque Resolution in Software Servo Systems Based on the Requirement of Control Performances, Trans. of the Institute of Electrical Engineers of Japan, vol. 114-C, no. 7/8, pp. 783-788, 1994. (in Japanese)

Chapter 5 [21][22]
[21] M. Nakamura, H. Yoshino, S. Goto and N. Kyura: A Method for Measurement of Torque Saturation Characteristic for Mechatronic Servo Systems, Proceedings of the 14th SICE Kyushu Branch Annual Conference, pp. 355-358, 1995. (in Japanese)
[22] S. Goto, M. Nakamura and N. Kyura: Trajectory Generation for Contour Control of Mechatronic Servo Systems Subjected to Torque Constraints, Proceedings of the 1994 Korean Automatic Control Conference, IS-04-3, pp. 66-70, 1994.

Chapter 5 [21][22]
[23] S. Goto, M. Nakamura and N. Kyura: Method for Modifying Taught Data for Accurate High Speed Positioning of Robot Arm, Trans. of SICE, vol. 27, no. 12, pp. 1396-1404, 1991. (in Japanese)
[24] S. Goto, M. Nakamura and N. Kyura: A Modified Taught Data Method by Using a Gaussian Network for Accurate Contour Control of Mechatronic Servo

Systems, Trans. of the Institute of Electric Engineers of Japan, vol. 115-C, no. 1, pp. 111-116, 1995. (in Japanese)
[25] S. Goto, M. Nakamura and N. Kyura: Accurate Contour Control of Mechatronic Servo Systems Using Gaussian Networks, IEEE Trans. Indust. Elect., vol. 43, no. 4, pp. 469-476, 1996.
[26] M. Nakamura, K. Tsukahara, S. Goto and N. Kyura: Contour Control of Flexible Manipulators by Use of Modified Taught Data Method, Trans. of SICE, vol. 33, no. 2, pp. 143-144, 1997. (in Japanese)
[27] T. Katayama: Basic of Feedback Control, Asakura Shoten, pp. 62-64, 1987.
[28] S. M. Shinners: Modern Control System Theory and Application, Massachusetts : Addison-Wesley, pp. 286-289, 1972.
[29] M. Kawato: Adaptation and Learning for Autokinetic Control, Journal of the Robotics Society of Japan, vol. 4, no. 2, pp. 184-193, 1986. (in Japanese)
[30] S. Lee and R. M. Kil: A Gaussian potential function network with hierarchically self-organizing learning, Neural Networks, vol. 4, pp. 207-224, 1991.
[31] B. Widrow and M. A. Lehr: 30 years of adaptive neural network: perceptron, madaline, and backpropagation, in C. Lau (Ed.), Neural Networks, New York, IEEE Press, Part 2, pp. 27-53, 1992.

Chapter 7 [32][33][34]
[32] S. Goto, M. Nakamura, S. Oka and N. Kyura: A Method of Synchronous Position Control for Multi Servo Systems by Using Inverse Dynamics of Slave Systems, Trans. of SICE, vol. 30, no. 6, pp. 669-676, 1994. (in Japanese)
[33] M. Nakamura, D. Hiyamizu, K. Nakamura and N. Kyura: A Method for Synchronous Position Control of Mechatronic Servo System with Master-Slave Axes by Use of Second Order Model, Trans. of SICE, vol. 33, no. 9, pp. 975-977, 1997. (in Japanese)
[34] M. Nakamura, D. Hiyamizu and N. Kyura: A Method for Precise Contour Control of Mechatronic Servo System with Master-Slave Axes by Use of Synchronous Position Control, Trans. of SICE, vol. 33, no. 4, pp. 274-279, 1997. (in Japanese)
[35] N. Kyura and Y. Hiraga: A Method of Following Control between Two Servo Systems, Bulletin of Japan Patent Office, Shou63-268011, 1988. (in Japanese)

Appendix
[36] N. Mizugami: Automatic Control, Asakura Shoten, pp. 23-41, 1968. (in Japanese)
[37] H. Kogou and T. Mita: Basic of System Control, Jikkyo Shuppan, pp. 124-130, 1979. (in Japanese)

Index

0th order hold, 57, 60
1st order model, 123, 144, 161, 162, 164, 166
1st order servo, 132, 133
1st order system, 39, 54, 70, 73, 128
2nd order model, 125, 133, 137
2nd order system, 32, 60, 81, 86, 87, 144
4th order model, 17, 20

A/D conversion, 94
acceleration output, 101
acceleration saturation property, 101
actual maximum acceleration output, 104
actuator, 39
allowable error, 36, 163
amplitude of angular velocity output deterioration, 93
amplitude of position fluctuation, 92
amplitude of position output deterioration, 93
analogue, 53
analogue servo system, 80
angular acceleration resolution, 87, 89, 93
angular velocity fluctuation, 90
approximation error, 42
axis resonance, 19
axis resonance filter, 19, 20

band pass filter, 143
bearing, 100
bit number, 94
Bode diagram, 130, 131

carrier frequency, 19
characteristic root , 25
characteristic roots equation, 56
chip mounter, 17
circle approximation, 109
clip, 99
closed-loop control system, 123
cogging torque, 69
complex conjugate root, 24
continuous oscillation, 23
contour control, 30
contour control method of master-slave synchronous positioning, 160
control performance , 26
coordinate transform, 37
corner part, 162
Coulomb friction, 99
counter, 80
counter-electromotive force, 99
counter-electromotive force compensation, 99
current control part, 18, 19
current detector, 18
current feedback, 94
current interruption, 98
current loop, 20
current reference, 86, 98
cut-off frequency, 19, 53, 56, 130
cut-off frequency condition, 55, 56

D/A conversion, 94
D/A converter, 57, 69, 85, 86
damping factor, 22, 23, 144, 146, 147
dead time, 53, 57

Index

dead zone, 19
design of servo controller, 17
detection noise, 81
determination method of servo parameter, 23
difference computation, 81
digital, 53
discrete time interval, 53
discretization, 57
disturbance, 150, 151
dynamics, 81, 121

empirical rule, 17
encoder, 19, 69, 80, 87
encoder resolution, 79, 82, 83
error back propagation learning, 141
extended command, 162

feedback gain, 124, 132
feedforward compensation, 151
feedforward control, 122, 132
flexible arm, 144–146
flexible mechanism, 144
fluctuation of ramp response, 93
fluctuation period, 92
follow, 30
following control, 129
following locus, 121
following trajectory, 122
fractional control, 62
frequency domain, 126, 130
friction, 19
friction torque, 100

gain property, 130
Gaussian function, 138
Gaussian unit, 138
gear ratio, 20, 39

impulse response, 103
industrial robot, 17
inertia matrix, 39
inertial moment, 20
infinitesimal, 43
initial parameter, 138
initial value, 153, 163, 164
integral (I) action, 19, 98
integrator, 132
interference, 19

intermediate unit, 138
inverse dynamics, 122, 132, 137
inverse kinematics, 39
inverse system, 137

Jacobian matrix, 74
joint coordinate, 19, 37, 73
joint linearized model, 39

kinematics, 38

Laplace transform, 20
learning, 140, 141
learning rate, 141, 143
linear function, 139
linear interval, 160, 161
linear model, 101
linear region, 99
liniarizable region, 140
locus error, 72, 75, 163–166
locus irregularity, 69, 70, 73, 74
loss function, 140, 141
low pass filter, 81, 83
low speed 1st order model, 31
low speed operation, 35

management part, 18
master-axis, 149
master-slave synchronous positioning control method, 149
mathematical model, 20
maximum acceleration, 116
maximum acceleration output, 104
maximum allowable current, 98
maximum phase, 131
maximum torque, 94, 129
maximum velocity, 30, 125
mean, 138
mechanism, 19, 30
mechanism part, 18, 100
mechatronic servo system, 17, 18
micro processor, 53
middle speed 2nd order model, 32
middle speed operation, 36
minimum order observer, 126
model construction, 17
model outputs error, 35
modeling error, 36, 135, 137, 140, 164, 165

modification element, 121, 125, 144, 145, 153, 161
modified taught data method , 121
module robot, 62
moment of inertia, 23
motor axis equivalent inertial moment, 23
motor part, 18

natural angular frequency, 22, 144, 146
natural frequency, 19
NC machine tool, 17
Neumann series, 68
nonlinear coordinate transform, 19
nonlinear term, 137
nonlinear transform, 38
normal vector, 72
normalized 4th order model, 22, 23, 31
numerical differential, 143
numerical integral, 143

objective joint angle, 37
objective locus, 121
objective trajectory, 37, 39
observation noise, 152
oscillation, 23
overload current, 98
overshoot, 23
overshoot condition, 54

P control, 20, 86
Pade approximation, 56
parallel link, 39
phase characteristics, 130, 131
phase-lead compensation, 132
PI control, 86
PI controller, 19
playback, 121
pole, 124, 128
pole of observer, 127, 128, 132
pole of regulator, 124, 128
pole of servo system, 132
position control part, 18
position detector, 18
position fluctuation, 90, 91
position loop, 62
position loop gain, 21, 31, 33
positioning control, 30
positioning error, 89, 93

positioning preciseness, 80, 88
positioning precision, 93
power amplifier, 19, 86
power amplifier part, 18
principal root, 24
proper, 122, 130, 132, 153
proportional constant, 150
pulse, 87
pulse counter, 19
pulse output, 80
pulse signal, 19

quantization error, 57, 86
quantization term, 87, 89

ramp input, 24, 30, 60
ramp response, 24, 31, 89
rated speed, 32, 39
rated torque, 98
reaction force, 20
real pole, 24
reduced order, 29
reduced order model, 29, 31
reference input generator, 18
reference input time interval, 40, 59, 69, 70, 75
resolution, 69, 80
resonance frequency of axis torsion, 19
response component, 24
rigid body system, 146
rigid connection , 22
rigid link, 38
robustness, 164

sampling control, 53, 57
sampling control system, 53
sampling frequency, 54, 56, 57
sampling time, 82
sampling time interval, 53, 54, 86
sampling time interval for velocity loop, 88
saturation region, 98, 105
saw tooth state cycle disturbance, 157, 159
self-organized robot, 62
semi-closed type control system, 122
sensor, 18
servo controller, 18–20
servo motor, 18
servo parameter, 22, 82

servo theory, 132
slave-axis, 149
small interval, 39
software servo, 79
software servo system, 80, 81, 86
spring constant, 20, 22
squared integral, 31
standard deviation, 138
state-space representation, 123, 125
steady state, 70
steady-state error, 70
steady-state value, 88
steady-state velocity deviation, 24, 31, 32
steady-state velocity fluctuation, 61
step disturbance, 154, 155
step-wise function, 71, 87
stick-slip, 69
structure, 138

tachogenerator, 57
tapping process work, 149
taught data, 121, 122
Taylor expansion, 42, 48, 74, 139
teaching playback robot, 122
teaching signal, 140, 141
theoretical acceleration output, 101
theoretical torque output, 103
time constant, 130
time domain, 129
torque, 21
torque command, 87
torque disturbance, 19
torque limitation, 129
torque of acceleration-deceleration, 99
torque quantization, 86, 87
torque quantization error, 88
torque resolution, 86, 93, 94
torque saturation , 97
torque saturation curve, 101, 104
torque saturation property , 100
total inertial moment, 22
tracking control method between two servo systems, 153, 154
trajectory speed, 30
transient state, 70

transient velocity fluctuation, 66
trapezoidal wave, 30
triangle inequality, 43, 48
two mass model, 20

undershoot, 56
unit, 138
unit step function, 71
unstable zero, 56

velocity amplifier gain, 21
velocity control part, 18
velocity controller, 20
velocity detection filter, 20
velocity detector, 18, 80
velocity disturbance, 162
velocity feedback, 81, 82
velocity fluctuation, 58, 62, 64, 82, 83
velocity fluctuation amplitude, 92
velocity fluctuation frequency, 83
velocity fluctuation period, 83
velocity fluctuation ratio, 83
velocity input reference, 150
velocity limitation, 125, 129
velocity loop, 19, 86, 126
velocity loop gain, 22, 33
velocity resolution, 82
velocity step input, 101
viscous friction, 99
viscous friction coefficient, 20, 22

weight of unit, 138
wind-up phenomenon, 98
working coordinate, 19, 37, 73, 123
working linearizable approximation possible region, 44
working linearized approximation error, 42–44
working linearized approximation trajectory, 41
working linearized model, 37
working precision, 109

zero, 128, 133

Lecture Notes in Control and Information Sciences

Edited by M. Thoma and M. Morari
2000–2004 Published Titles:

Vol. 250: Corke, P.; Trevelyan, J. (Eds)
Experimental Robotics VI
552 p. 2000 [1-85233-210-7]

Vol. 251: van der Schaft, A.; Schumacher, J.
An Introduction to Hybrid Dynamical Systems
192 p. 2000 [1-85233-233-6]

Vol. 252: Salapaka, M.V.; Dahleh, M.
Multiple Objective Control Synthesis
192 p. 2000 [1-85233-256-5]

Vol. 253: Elzer, P.F.; Kluwe, R.H.; Boussoffara, B.
Human Error and System Design and Management
240 p. 2000 [1-85233-234-4]

Vol. 254: Hammer, B.
Learning with Recurrent Neural Networks
160 p. 2000 [1-85233-343-X]

Vol. 255: Leonessa, A.; Haddad, W.H.; Chellaboina V.
Hierarchical Nonlinear Switching Control Design with Applications to Propulsion Systems
152 p. 2000 [1-85233-335-9]

Vol. 256: Zerz, E.
Topics in Multidimensional Linear Systems Theory
176 p. 2000 [1-85233-336-7]

Vol. 257: Moallem, M.; Patel, R.V.; Khorasani, K.
Flexible-link Robot Manipulators
176 p. 2001 [1-85233-333-2]

Vol. 258: Isidori, A.; Lamnabhi-Lagarrigue, F.;
Respondek, W. (Eds)
Nonlinear Control in the Year 2000 Volume 1
616 p. 2001 [1-85233-363-4]

Vol. 259: Isidori, A.; Lamnabhi-Lagarrigue, F.;
Respondek, W. (Eds)
Nonlinear Control in the Year 2000 Volume 2
640 p. 2001 [1-85233-364-2]

Vol. 260: Kugi, A.
Non-linear Control Based on Physical Models
192 p. 2001 [1-85233-329-4]

Vol. 261: Talebi, H.A.; Patel, R.V.; Khorasani, K.
Control of Flexible-link Manipulators Using Neural Networks
168 p. 2001 [1-85233-409-6]

Vol. 262: Dixon, W.; Dawson, D.M.; Zergeroglu, E.; Behal, A.
Nonlinear Control of Wheeled Mobile Robots
216 p. 2001 [1-85233-414-2]

Vol. 263: Galkowski, K.
State-space Realization of Linear 2-D Systems with Extensions to the General nD ($n>2$) Case
248 p. 2001 [1-85233-410-X]

Vol. 264: Baños, A.; Lamnabhi-Lagarrigue, F.;
Montoya, F.J
Advances in the Control of Nonlinear Systems
344 p. 2001 [1-85233-378-2]

Vol. 265: Ichikawa, A.; Katayama, H.
Linear Time Varying Systems and Sampled-data Systems
376 p. 2001 [1-85233-439-8]

Vol. 266: Stramigioli, S.
Modeling and IPC Control of Interactive Mechanical Systems – A Coordinate-free Approach
296 p. 2001 [1-85233-395-2]

Vol. 267: Bacciotti, A.; Rosier, L.
Liapunov Functions and Stability in Control Theory
224 p. 2001 [1-85233-419-3]

Vol. 268: Moheimani, S.O.R. (Ed)
Perspectives in Robust Control
390 p. 2001 [1-85233-452-5]

Vol. 269: Niculescu, S.-I.
Delay Effects on Stability
400 p. 2001 [1-85233-291-316]

Vol. 270: Nicosia, S. et al.
RAMSETE
294 p. 2001 [3-540-42090-8]

Vol. 271: Rus, D.; Singh, S.
Experimental Robotics VII
585 p. 2001 [3-540-42104-1]

Vol. 272: Yang, T.
Impulsive Control Theory
363 p. 2001 [3-540-42296-X]

Vol. 273: Colonius, F.; Grüne, L. (Eds)
Dynamics, Bifurcations, and Control
312 p. 2002 [3-540-42560-9]

Vol. 274: Yu, X.; Xu, J.-X. (Eds)
Variable Structure Systems:
Towards the 21st Century
420 p. 2002 [3-540-42965-4]

Vol. 275: Ishii, H.; Francis, B.A.
Limited Data Rate in Control Systems with Networks
171 p. 2002 [3-540-43237-X]

Vol. 276: Bubnicki, Z.
Uncertain Logics, Variables and Systems
142 p. 2002 [3-540-43235-3]

Vol. 277: Sasane, A.
Hankel Norm Approximation for Infinite-Dimensional Systems
150 p. 2002 [3-540-43327-9]

Vol. 278: Chunling D. and Lihua X. (Eds)
H_∞ Control and Filtering of
Two-dimensional Systems
161 p. 2002 [3-540-43329-5]

Vol. 279: Engell, S.; Frehse, G.; Schnieder, E. (Eds)
Modelling, Analysis, and Design of Hybrid Systems
516 p. 2002 [3-540-43812-2]

Vol. 280: Pasik-Duncan, B. (Ed)
Stochastic Theory and Control
564 p. 2002 [3-540-43777-0]

Vol. 281: Zinober A.; Owens D. (Eds)
Nonlinear and Adaptive Control
416 p. 2002 [3-540-43240-X]

Vol. 282: Schröder, J.
Modelling, State Observation and
Diagnosis of Quantised Systems
368 p. 2003 [3-540-44075-5]

Vol. 283: Fielding, Ch. et al. (Eds)
Advanced Techniques for Clearance of
Flight Control Laws
480 p. 2003 [3-540-44054-2]

Vol. 284: Johansson, M.
Piecewise Linear Control Systems
216 p. 2003 [3-540-44124-7]

Vol. 285: Wang, Q.-G.
Decoupling Control
373 p. 2003 [3-540-44128-X]

Vol. 286: Rantzer, A. and Byrnes C.I. (Eds)
Directions in Mathematical Systems
Theory and Optimization
399 p. 2003 [3-540-00065-8]

Vol. 287: Mahmoud, M.M.; Jiang, J. and Zhang, Y.
Active Fault Tolerant Control Systems
239 p. 2003 [3-540-00318-5]

Vol. 288: Taware, A. and Tao, G.
Control of Sandwich Nonlinear Systems
393 p. 2003 [3-540-44115-8]

Vol. 289: Giarré, L. and Bamieh, B.
Multidisciplinary Research in Control
237 p. 2003 [3-540-00917-5]

Vol. 290: Borrelli, F.
Constrained Optimal Control
of Linear and Hybrid Systems
237 p. 2003 [3-540-00257-X]

Vol. 291: Xu, J.-X. and Tan, Y.
Linear and Nonlinear Iterative Learning Control
189 p. 2003 [3-540-40173-3]

Vol. 292: Chen, G. and Yu, X.
Chaos Control
380 p. 2003 [3-540-40405-8]

Vol. 293: Chen, G. and Hill, D.J.
Bifurcation Control
320 p. 2003 [3-540-40341-8]

Vol. 294: Benvenuti, L.; De Santis, A.; Farina, L. (Eds)
Positive Systems: Theory and Applications (POSTA 2003)
414 p. 2003 [3-540-40342-6]

Vol. 295: Kang, W.; Xiao, M.; Borges, C. (Eds)
New Trends in Nonlinear Dynamics and Control,
and their Applications
365 p. 2003 [3-540-10474-0]

Vol. 296: Matsuo, T.; Hasegawa, Y.
Realization Theory of Discrete-Time Dynamical Systems
235 p. 2003 [3-540-40675-1]

Vol. 297: Damm, T.
Rational Matrix Equations in Stochastic Control
219 p. 2004 [3-540-20516-0]

Vol. 298: Choi, Y.; Chung, W.K.
PID Trajectory Tracking Control for Mechanical Systems
127 p. 2004 [3-540-20567-5]

Vol. 299: Tarn, T.-J.; Chen, S.-B.; Zhou, C. (Eds.)
Robotic Welding, Intelligence and Automation
214 p. 2004 [3-540-20804-6]

Vol. 301: de Queiroz, M.; Malisoff, M.; Wolenski, P. (Eds.)
Optimal Control, Stabilization and Nonsmooth Analysis
373 p. 2004 [3-540-21330-9]

Vol. 302: Filatov, N.M.; Unbehauen, H.
Adaptive Dual Control: Theory and Applications
237 p. 2004 [3-540-21373-2]

Vol. 303: Mahmoud, M.S.
Resilient Control of Uncertain Dynamical Systems
278 p. 2004 [3-540-21351-1]

Vol. 304: Margaris, N.I.
Theory of the Non-linear Analog Phase Locked Loop
303 p. 2004 [3-540-21339-2]

Vol. 305: Nebylov, A.
Ensuring Control Accuracy
256 p. 2004 [3-540-21876-9]

Vol. 306: Bien, Z.Z.; Stefanov, D. (Eds.)
Advances in Rehabilitation Robotics
472 p. 2004 [3-540-21986-2]

Printing and Binding: Strauss GmbH, Mörlenbach